Statistical Testing with jamovi

Sport

Dedicated to The Bystanders Society

Statistical Testing with jamovi Sport

SECOND EDITION

Statistics without Mathematics series

General Editor
Cole Davis

Vor Press

© Cole Davis 2023
Second edition

First published in Great Britain in 2020

by Vor Press

21 Chalk Hill Road,

Norwich NR1 1SL

www.vorpress.com

This book has been deposited with the British Library.

paperback
ISBN 978-1-915500-22-9

hardcover
ISBN 978-1-915500-21-2

ebook
ISBN 978-1-915500-23-6

Contents

Statistical testing fundamentals

Part 3

Part 5

Bayesian statistics
introduced   262

Part 1

Background knowledge

Chapter 1 – Introduction

Is this book for me?

With worked examples covering different sports, this book will appeal in particular to students. Tutors will also find this of use, both for its advanced tests and also its introduction to Bayesian statistics. Those involved in applied research should also be able to use this book profitably.

Short and without any mathematical formulae, the book starts with the assumption that you know nothing about statistics and builds you up slowly rather than hurling a comprehensive toolkit at you. But toolkit it is. By the end of Part 2, you should be able to analyze a wide range of common problems. And then we take you further.

I'm interested in sport and they make me do statistics

Apart from being made to do it, I can think of some good reasons for studying statistics. Practically speaking, you can not only find out if your ideas work in the real world but also demonstrate the extent of their success. It is also likely to stand you in good stead in employment. Whether you are in sports management or coaching, or just want to improve performance, if you are known to be a confident user of statistics, you become more of an attractive proposition. You will be able to evaluate techniques so that managers and coaches can get a better idea of what works and how; interpret surveys so that organizations gain insights into what the public or sports people are thinking; and integrate observations, interviews and focus groups into more obviously quantifiable results.

You might even find, perhaps later in life, that you enjoy research and data analysis and end up as an information specialist. There is a demand for people with analytical skills.

Is it really relevant to sport?

From the mid-nineteenth century, individual and team performances in baseball were represented statistically using the 'box score'. Between the mid and late twentieth century, baseball statistics, eventually given the snazzy title 'sabermetrics', became the source of controversy at a national level. In particular, different conditions were investigated to see what difference they made to performance. Basketball, another high-scoring game, also came under the number-crunchers' microscope. British sport, with its rather messier goals and tries, took rather longer to warm to statistics, but sport (and gambling) found new ways of examining performance, with quantitative analysis becoming quite an industry by the early twenty-first century (Kucharski 2016).

All the mathematics you will need

This book is for people who need to use statistical tests but who do not feel comfortable with mathematical explanations. To those who say that only equations express statistical concepts well, I provide the answer given by the publisher of another of my books: "This is for the rest of us" (thanks, Ted). There is no assumption of prior statistical knowledge and no use of formulae. You need no mathematical knowledge other than basic arithmetic, as applied to the decimal system. If you understand the following, you are ready to go:

Differences in sizes within the decimal system; these examples gradually decrease in size, towards zero:

1 .75 .5 .25 .125 .1 .07 .05 .03 .01 .005 .001

Beyond zero, we see negative numbers, with increasing negativity away from zero:

- 0.001 - 0.002 - .005 - .01 -.02 -.05 -.07 -.1 -.125 -.5 -.75 -1 -2

The sign < means 'less than', for example: .005 < .01. The sign > means 'greater than', for example .02 > .01. We use <= to mean 'less than or equal to'; we use >= to mean 'greater than or equal to'.

To multiply a number by itself is to 'square' a number. For example, 0.02×0.02, or '0.02 squared', equals 0.0004, a much smaller number. Note that if you multiply two negatives, for example, $-.02 \times -.02$, this should also yield a positive number (here, .0004). Try out your calculator to make sure that it can calculate the double negative correctly. If it doesn't, remove both negatives from your calculations.

In some instances, you will see * for multiplication, e.g. 4 * .05 = 0.2.

That is it. You have enough mathematical knowledge to use this book. You will come into contact with some statistical terms such as mean, median, mode and variance, but even those won't require any formulae or mathematical jiggery-pokery.

What is the teaching strategy of this book?

As well as refusing to use formulae, I have tried to make this book as short as is reasonably possible. I know that there is a feeling of security in having a 1200 page tome that appears to have everything. It is my belief, however, that instructions which convey the most important information are more likely to be understood and remembered. Too much information and the reader is not sure what is of real value. In short, this is not a course on statistics; it is a practical book on statistical testing.

The content is also geared to learning needs. Recognizing that people can only take in so much at one time, the earlier worked examples in any given chapter are simpler, with more information accreting later on. The decision to include reporting at the end of Part 2 reflects the likelihood of having to make presentations while studying. More technical debates and over-arching material occur later in the book.

Worked examples are provided for each test. The usual habit of posing questions at the end of each chapter has been eschewed. This is primarily because the real world does not hold your hand and say "Hm, this looks like a job for a Kruskal-Wallis test", but also because there seems little reason to keep the reader under test conditions at all times. On the other hand, to ensure understanding at the level of being able to choose the right tests for basic problems, exercises in the form of simple case studies are available near the end of Part 2.

Several of the controversies in the world of statistical testing are discussed in passing. Apart from any entertainment value, this is in my opinion quite necessary. Authors who skirt around awkward choices are not around when the test user is thinking 'well is it me, or does this seem rather strange?' In many cases, the fool is not the test user, who is merely stumbling upon the nub of a debate that has been raging for years outside the cosy world of textbooks.

What's in the book?

Each part has a particular purpose. The rest of this Part 1 provides the basic underpinnings for carrying out data analysis.

Part 2 is devoted to the statistics that have always been taught at undergraduate level. For many readers, this will be sufficient for quite a long time. Chapter 5 deals with analyses of differences between variables; are the differences in a study likely to be generalizable to the wider world or are they likely to be a fluke? Chapter 6 deals with relationships between variables. The later part of the chapter includes multiple regression, a technique which used to be neglected in introductory textbooks but is now increasingly used. If you've always wondered about whether or not some of those interviews and focus groups could be quantified in some way, then Chapter 7 is for you. It counts observations, analyzing the frequencies for each category. Most introductory books only cover 'Chi Squared'; this book goes further, as does Jamovi. Exercises then follow to test your general understanding of which tests are usable in what contexts. There is also a chapter on reporting. This covers both written work and verbal presentation, although it does not teach how to write academic reports: apart from the multitude of works on the subject, this decision reflects the fact that most universities, let alone countries, have rather different expectations as to what exactly should be included in a formal report.

Part 3 is primarily about extensions to ANOVA: factorial ANOVA, ANCOVA and MANOVA.

Part 4 is in advanced territory. Factor analysis and principal components analysis are likely to turn up in postgraduate projects. Logistic regression provides both the flexibility to use varied types of data and the ability to hone in on findings to examine the extent of their accuracy and predictive capabilities.

Then come some features which are less usual in introductory statistics. The brief chapter on partial and semi-partial correlations allows the test user to hone in on the relationships between specific variables of interest. Then comes an introduction to Bayesian statistics, an alternative to the 'significant or not' dichotomy of classical statistics.
*

The final section contains chapters on two specialist methods for exploring data visually. The Kaplan-Meier curve is a widely applicable survival analysis tool which examines events over time. Cluster analysis focuses on groups of cases and is of particular use in market research, web analytics and finding common factors amongst sports and other professionals.

---

*A note on Bayes: Please don't complain if your tutors do not use Bayesian tests when teaching you. This type of statistical analysis has only become computable in the 21st century and is not yet in the mainstream – you're ahead of the game!

How much do I need to read? Subtitle: How much can I skip?

Beginners

You need to read all of the first part. Ideally you should read all of the second part, following all of the worked examples and completing the chapter of exercises. Then you can move on to any advanced chapters as and when you need them or they take your interest (no, really?). I would suggest also looking at the chapter on reporting, especially if somebody wants you to get up and address an audience.

If you are in a hurry to use an advanced test

If you don't wish to read all of Part 2 on testing fundamentals, do cover the necessary groundwork:

Within Chapter 5, if you want to use tests of differences using more than two conditions (analyses of variance, Friedman, Kruskal-Wallis), you really should study the two-condition tests first (T-Tests, Wilcoxon, Mann-Whitney) – it won't take long. Factorial ANOVA should of course be preceded by one-way ('univariate') ANOVA. If you are considering the use of MANOVA, you should have already mastered Chapter 5 and Chapter 10 on factorial ANOVA. Within Chapter 6, correlations should precede regression, and both are necessary before undertaking multiple regression. Log-linear analysis should be preceded by the earlier parts of Chapter 7, particularly the chi squared test of association, and also Chapter 6, especially the section on hierarchical regression. ANCOVA should be preceded by Chapter 5 and Chapter 6 (and still avoided, perhaps). Factor analysis and/or principal components analysis should be preceded by the chapter on correlations and regression. Logistic regression should be preceded by Chapter 6 in its entirety. Partial correlations should be read on top of the part of Chapter 6 dealing with correlations.

A really basic course for beginners

You could reduce your reading in order to cover traditional introductory topics only. Read all of Part 1 and then Chapter 5. Within Chapter 6, ignore hierarchical and sequential regression.

In Chapter 7, you could read just the introduction and the section on the chi square Test of Association, although including the binomial and multinomial (chi squared Goodness of Fit) tests provides a broader overview of categorical analysis. Although a logical progression, log-linear analysis is probably unnecessary for beginners. You can omit the McNemar test.

I would recommend doing the exercises in Chapter 8, skimming Chapter 9 on reporting research, and reading Chapter 10 on factorial ANOVA and multiple comparisons.

Intermediates and returners

You can skip the first part, although I would recommend Chapter 4, on null hypothesis testing, especially on the subject of p values and critical values. If it was all a long time ago, the chapter on descriptive statistics might be helpful.

The part devoted to statistical fundamentals could be used selectively, dependent on need, but it would not take long to skim through the chapter just to become familiar with how Jamovi works (similar in feel to SPSS but with simplified setting-up procedures and reduced statistical output).

If there are advanced tests that you have not used before, follow the advice for beginners, 'If you are in a hurry'. Obviously, one should build upon accumulated knowledge.

Why use Jamovi?

It is free and open source. While data input is similar to SPSS, its expensive rival, procedures have been streamlined – note in particular the comparative simplicity of setting up a repeated measures ANOVA – and options can be altered with ease. Jamovi also contains effect sizes and confidence intervals, of increasing importance in modern statistics.

Jamovi has a core of tests which are usable within a typical under-graduate course, but the package offers some additional tests which will offer a broader learning experience. One key focus of Jamovi is reproducibility; options chosen by the user may be saved within a file. Annotation is built in, and Jamovi has developed its own editing suite.

A brief note on data entry

Jamovi previously only used files with the .csv suffix (comma separated values); these represent the active tab within a conventional spreadsheet (File/Save As/ save as type) and are used in the worked examples in this book. However, the package can now read Excel (.xlsx) and LibreOffice (.odt) files, as well as file formats from a range of statistical packages. As well as opening data sets from files, it is also possible to type in data directly. Here, I open the file Basic.csv, with just 10 cases (files can be downloaded from the publisher's website).

case	salary	parentIncome	edStatus	parentEdStatus	gender	ethnic	manager	grade
1	24	20	4	3	female	white	middle	coach
2	30	28	5	5	female	asian	junior	trainer
3	15	12	3	3	male	black	junior	trainer
4	50	27	5	4	female	black	senior	manager
5	28	34	4	5	male	white	middle	coach
6	20	21	4	3	female	white	junior	trainer
7	22	20	2	2	male	white	junior	trainer
8	45	36	5	5	male	asian	senior	manager
9	38	28	1	2	male	white	middle	coach
10	27	35	4	4	female	white	middle	coach

Each observation should have its own row. In a same-subjects design, say looking at the relationship between salary and parental income, the relevant variables are used in a straightforward manner. In a between-subject design, such as examining differences in educational background among managerial groups, then a grouping variable such as 'manager' is used to examine a metric such as 'edStatus'. This will become very familiar to you as you read the book, which has worked examples for each test.

Details about data handling may be found on the Jamovi website.

Main contributors

Cole Davis. Elena Rychkova. Marianne Vitug.

Acknowledgements

Other contributors and advisers at different times in the gestation of the series have included Michy Alice, Ed Boone, Iain Buchan, Winston Chang, Michael Fay, Henriette Hogh, Jonathon Love, Sharon McGrayne, Abraham Mathew, Richard Morey, Ofra Reuven, William Revelle, Grant Schneider, E-J Wagenmakers and Douglas Wolfe. I am indebted to them for their advice, encouragement and cooperation.

I am also grateful to previous commissioning editors who have provided useful criticism and support for my ideas on teaching statistics to complete strangers. In approximate chronological order, these were Karen Winter (née Bowler) at Policy Press, with John Manger and Ted Hamilton at CSIRO. Also, I appreciated the stimulation from discussions with their counterparts at John Wiley and Routledge.

The additional assistance and further guidance of Ted Hamilton, publisher emeritus, is deeply appreciated. Also, step forward, or rather, stay sitting, Sophie Hamilton, who graces the front cover of this book.

I would also like to thank Casper Albers of Groningen University and Boris Mayer of the University of Bern for their ideas on incorporating Bayesian statistics into a general introductory textbook.

The British Library was a source of limitless materials and those other priceless resources, peace and quiet. I would like to thank the Society of Authors for their continued support in a variety of ways.

The assistance and hospitality of David and Sue Weisberg have been an enduring feature.

I apologize if I've missed anybody or failed to take up advice as proffered. Any errors or questionable judgements are my own.

Second edition

This edition includes a new chapter on cluster analysis, a revised chapter on partial correlations, and refers specifically to Jamovi.

Feedback request

If you wish to submit reviews, make suggestions for improvements or point out errors, please contact me via the publishing website.

Cole Davis 2023

Chapter 2 – Research design

Experiments, control groups, variables and other terms

There are plenty of books on research design, but this book is primarily about statistical testing, particularly using Jamovi. This being the case, we will only consider the basics, as they pertain to testing. As you go through the examples in the book, these terms will become more familiar and easier to understand.

The terms **independent variable** and **dependent variable** will be referred to regularly. These are respectively the variables being manipulated and the variables affected by such manipulation. You will also come across the terms **predictor** and **criterion**, the former variable influencing the situation and latter variable being the item of measurement. The two pairs of terms are interchangeable in much of the literature. Strictly speaking, **experiments** should refer to independent and dependent variables, as shown in this example:

One group of coaches, randomly chosen, is given additional training in how to provide feedback to their gymnasts; this is the **experimental group**. Another group of coaches is not given training; this is the **control group**.

The difference between the two **conditions** (in data analysis, often referred to as **levels**) is judged by levels of errors over the next quarter, or whatever measure is considered most suitable. The independent variable (sometimes abbreviated to **iv**) is the existence or otherwise of feedback training; the dependent variable (**dv**) is performance.

We could have a **quasi-experiment**, perhaps using past records to see how feedback affects performance, or maybe using naturally occurring groups – for example, one sporting club has this training and a similar club does not – and we contrast the groups. Here we do not really have a control group. Strictly speaking, we should not use the terms 'independent variable' or 'dependent variable'. The use or otherwise of training is the predictor, with the different levels of performance as the criterion (or 'measure'). However, you will see both sets of terms used in the literature, regardless of the official status of the research.

Generally speaking, however, we use the terms predictors and criteria in regression. For example, we may be interested in the effect of different pool temperatures and chlorine levels on swimmers' performances over a long distance. The different temperature and chlorine levels may be the predictors, whereas times and maybe physiological measures, if there is a clear linear relationship with the predictors, would be the criteria.

What we are trying to achieve

Some of the time we are trying to look at *differences* between conditions. This will be seen in particular with *t* tests, ANOVA, and their non-parametric equivalents. There is a further sub-divide, whether or not you use a **between subjects** design, also known as an **unrelated design**, where different people are tested in each condition, or a **same-subjects** design (also known as **within subjects** or **related design**). The advantage of using the same subject under the different conditions is that the **effect** under study is unlikely to be conflated with individual differences. Often, where we cannot use the same person, we may try to **pair** (or **match**) the subjects (participants, if human beings) in the relevant areas; for example, if we wanted to contrast two types of training method, but wanted each individual to experience only one of them, we could choose sportspeople to be paired according to sport, similar levels of tested intelligence, age, weight and gender, making them similar for the purpose of the experiment. **Paired tests** are used for both paired and same-subjects research. For obvious reasons, the same number of subjects is required in each condition.

Sometimes we are unable to use the same subjects. Perhaps the study would be adversely affected by participants experiencing more than one condition, or different subjects are the whole point of a study (males and females, in a gender study, are usually different). Then **unpaired tests**, known in Jamovi as **independent samples** tests, try to take into account individual differences. There may be different numbers of subjects in each condition.

At other times, we are interested in the *relationships* between conditions. In particular, we examine **correlations**, linear relationships between conditions, positive or negative. For example, we may study a range of different attributes to see if they are inter-related. Perhaps the more outward-going the person, the more supporters that he or she has. It should be noted that correlations do not necessarily demonstrate cause and effect. For example, it might be that having more supporters builds up a person's confidence and makes them more outward-going.

A correlation requires the intersection of informational pairs, for example with each individual's scores on one measure matched with their scores on another. For this reason, there must be equal numbers of subjects in each condition.

Data types

Another issue is the nature of the data. The above situations generally require measurable data, but the choice of test depends upon the granularity of the data. **Continuous** data is proportional and takes on a 'natural' feel, like a range of body weights, times and ages, 48, 50, 53, 55, 55, 56, 58 and so on. **Ordinal** data can include the results of an uncalibrated Likert scale (1 to 5, 1 to 7, 1 to 9 and so on) or rather coarsely grained, 'lumpy' data, such as 2, 6, 55, 55, 109. The implications will be discussed in Chapter 3.

Other data are counts of observations or incidents: **frequency**. For example, you may want to look at the incidence of horses falling at a particular fence over a particular period. You could find 30 cases of discrimination against members of staff working at a sports center, 10 such cases in recruitment, and 5 in dealing with members of the public, and so on. As well as such differences, you may have matrices of cases allowing a study of relationships. If we have also classified the cases into the different social classes of the victims, we could examine the relationship between the class of the victim and the type of discrimination committed against them.

Frequency counts - 10 personal trainers, 22 footballers, 7 coaches - are **categorical** data (also known as **nominal** or even **qualitative** data). There are no quantitative comparisons such as averages; all differences are qualitative. Like elephants and lamp posts, you don't usually consider each individual on the same scale; elephant number 3 and lamp post number 3 cannot be considered for their relative luminosity or suitability for climbing (I think).

Chapter 3 – Descriptive statistics

Central tendency

Given a set of figures, whether or not we compare it with another set of figures, we need some way of representing it. We may want the maximum and minimum values, but when it comes to statistical testing, we are usually more interested in central tendency, basically a representative value which is deemed to be typical. The measure of central tendency is usually the **mean**, the **median** or **mode**.

If we take this very small data set, 2 3 3 4 8, we can demonstrate the differences between the three measures of central tendency. Here we work them out by hand; we won't need to do it again!

The mean, usually what is meant by 'average', is calculated by dividing the sum of the variable by the total number of cases. The sum here is 20, the number of cases 5, so 20/5 equals a mean of 4.

The median is the mid-point in the range. Calculating the median requires moving to the outer limits, eliminating the values there and continuing until we reach the middle. So in this example, we first eliminate 2 on the left and 8 on the right, then rule out 3 on the left and 4 on the right. Our median is 3. (The even-numbered set 2 3 3 4 4 8 would have 3.5 as its median; eliminate 2 and 8, then 3 and 4, leaving 3 and 4 in the middle.)

The mode is the most common response. The mode for 2 3 3 4 8 is of course 3, the most common number. It is possible to have multiple mode values.

In practical terms, these measures have varying utility. Let us say that we are considering salaries. The strength of using the mean is that it takes into account everybody's income, from the stratospherically well-paid to the lowliest wage-earner; on the other hand, this can be a weakness, say if two or three billionaires distort our figures. The mode may counteract this effect, as it tells us the income of the largest number of people, perhaps of administrative workers; but this is hardly representative of the workforce as a whole. The median gives us a central value, perhaps that of a middle manager; it is useful, but does not take into account the number of people who are wealthy or poor. At this stage, I will merely say that the example shows the need to adapt interpretations to context; there is no magic button that just tells you everything.

As button-pressing is nevertheless quite enjoyable, I suggest opening the Basic.csv file and pressing Jamovi's Exploration tab, then clicking Descriptives in order to look at some descriptive statistics. Transfer the four numerical variables from the left to the Variables box.

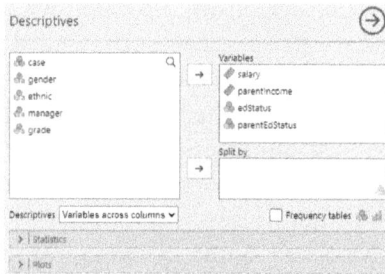

On the right-hand side is the display screen, where we see the default options:

Descriptives

	salary	parentIncome	edStatus	parentEdStatus
N	10	10	10	10
Missing	0	0	0	0
Mean	29.9	26.1	3.70	3.60
Median	27.5	27.5	4.00	3.50
Minimum	15.0	12.0	1	2
Maximum	50.0	36.0	5	5

As the Likert scales for edStatus and parentEdStatus have 5 points, the median figures reflect this, with a rather blunt 4 and 3.5; although the median is theoretically the best figure when dealing with non-parametric tests (of which, more later), it is often better to use the mean because of a more easily distinguishable spread of values.

In the Statistics section, you will also find such options as the mode and sum, which adds up all of the scores. You can also place a grouping variable into the Split by box to see the differences; in the case of 'grade', this means seeing the scores for different levels.

Dispersion

Dispersion is how the data is spread. By default, Jamovi shows us the minimum and maximum values, which are useful for checking for outliers and errors in data entry, and the standard deviation, which shows how spread out the data is; see the distribution chart a little later. The **range** is the difference between the minimum and maximum.

We probably do not need to see the variance, but the concept of **variance** is important. Many of the tests that we will use are based on the variability of statistics around the measure of central tendency. Parametric tests are particularly concerned with variance about the mean.

Assumptions for parametric tests

In general, parametric tests are considered to be more powerful than non-parametric tests and are therefore the weapons of first resort. On the other hand, if the data set does not meet the assumptions for parametric tests, parametric tests may come to conclusions about an imaginary data set; in such a case, the non-parametric test is preferred.

A non-parametric test ranks the data, as if they were all ordinal, and makes no assumptions about the distribution.

Unfortunately for test users, there are disagreements among statisticians about the extent to which these assumptions are necessary (in this book, you will find that this is not the only instance of mathematically inspired fisticuffs). More traditional test users insist on strict adherence to the assumptions for using parametric tests. Others note the robustness of parametric tests and are inclined to be rather less stringent about the assumptions.

We will consider each assumption in turn, looking first at the traditionally taught view and then some rather more relaxed practices.
Before you get too worried about this, you will find that employers and academics will have their own views on the subject, so be prepared to render unto Caesar.. [*]

One assumption is that the data are continuous. There is general agreement that the most lumpy data sets – stuff like 2 8 7 16 316 32 96 – are really unsuited to parametric tests (although they are common in many research studies). Beyond this, agreement tends to go out of the window. The more conservative test user asks, "if you halved the data, would the new value really be 50% of the old?" Think about, for example, the notion of a service being 'Ok', 'Average', 'Good', 'Very Good' and 'Superb' on a scale from 1 to 5; would halving Very Good mean Average?

When using a Likert scale, the conservative user would subject it to a parametric test only if the scale had already been calibrated in a pilot project. We do not have room to discuss calibration in this book, but you would do well to look up 'item response theory'. More relaxed test users use parametric tests on Likert scale results in all circumstances. [*]

Another assumption is homogeneity of variance. If you have one set of data that looks like 3 4 7 9 and another set of data that looks like 32 38

---

[*]My own take on non-parametrics is that when using data considered suitable for parametric tests, they generally show the same results. When the data are unsuitable, non-parametric tests may produce a more conservative result, with good reason.

[*]Pilot projects, by the way, are always advisable, as they give you the chance to iron out unexpected problems.

52 67, then parametric tests are not suited to working with them both together.

A third assumption is a normal (also known as Gaussian) distribution.

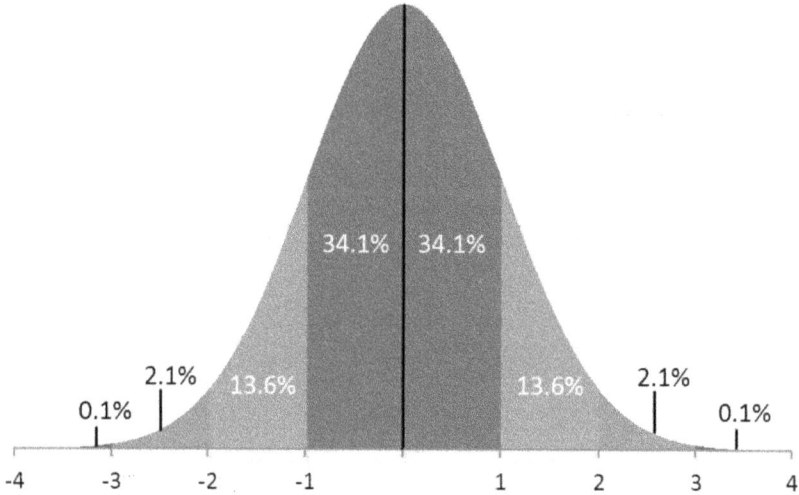

This is an idealized version of the normal distribution. Most of the data is near the central tendency, with less and less of the data lying at the extremes. The measures of dispersion on the axis at the bottom of the chart are standard deviations.

The implication is that your sample is representative of a population, an important concept when we consider null hypothesis testing in Chapter 4. Do note that a population does not have to be a population in the sense of all the people in a geographical area (although it can be). The population may be sports science students in the USA, the unemployed in London or libraries in Australia; the sample is assumed to be representative of that population.

Testing for distribution

In order to decide whether or not we have a distribution which is suitable for parametric test usage, we usually use an heuristic, a practical method which while not guaranteed to work, usually does the trick.

With reasonably large data sets, you can use the histogram and density charts within the Plots section to examine the distribution. Small data sets are unlikely to get an approximation of the normal curve shown above. This leads some traditionalists to recommend non-parametric tests for small data sets, although others will live with parametric tests if they meet fairly strict assumptions.

Returning to the Statistics section within Jamovi's Descriptives, go to the Distribution options and select Skewness and Kurtosis. Skewness is how far the data is spread away from the mean in one direction or the other. Kurtosis is more to do with the shape of the curve, in particular the weight of its 'tails'. For samples of less than 50, consider Skewness and Kurtosis readings of within 1.96 (or -1.96) to be acceptable; although more liberal readings go from +2 to -2. For 50 to 300 cases, the limit should be 3.29 (Kim 2013). For samples larger than 300, use the Plots option within Descriptives, also accepting a maximum of 2 for Skewness and 7 for Kurtosis (West *et al* 1996).

A more modern approach to testing for normality is to use the Shapiro-Wilk test (Razali and Wah 2011). One way of using this is to use the Normality assumption check within the One-Sample T-Test as shown in Chapter 4, or within Jamovi's Descriptives. A significant result using this test indicates a pattern that is probably not from a normal distribution. So you want a non-significant result from the Shapiro-Wilk to continue using a parametric test. We turn to the notion of significance, or perhaps more strictly, non-significance, in Chapter 4.

Chapter 4 – Null hypothesis significance testing

Null hypothesis significance testing (NHST) particularly features in classical statistics (also referred to as 'frequentist' statistics by Bayesians, who turn up at the end of the book). As it is the most commonly taught approach, it is particularly important to understand it.

I will keep this fairly short; your part of the deal is to return to this chapter whenever you feel less than clear about the subject. However, you should get more of a feel for it when you have gone through some of the worked examples in the book.

Let us say that the overall player ratings for tennis training center A are higher than the ratings for center B, but we are not sure if the difference (the **effect**) is a meaningful one. Or perhaps performance appears to be related to funding issues, but we want to know whether or not the relationship (again, the effect) is merely subject to chance.

In NHST, the statistical test examines the **sample** – the cases for which we have evidence – and considers it within the likely **population** from which the sample is drawn. This is why this area of statistics is often referred to as **inferential statistics**: it makes inferences from the data in the sample about the population as a whole. So we may study the behavior of a sample of 30 sports science students, assuming that they are reasonably representative of the population, perhaps sports science students nationwide.

The default position assumed by a statistical test is the **null hypothesis**, sometimes represented as H0. The null hypothesis is that the findings do not differ significantly from chance, noise or experimental error. More prosaically, the null hypothesis says by default that your beloved effect is just garbage. The test's essential role is to tell you whether or not it is reasonable to reject the null hypothesis; that your effect is not random variance. A scientific principle is being maintained, that of falsifiability: according to Popper (1968), in science one can only falsify a theory. We can find out if the null hypothesis can be upheld, yes, but we can't make a direct claim for an effect.

The hypothesis that you are trying to prove in your study is called the '**alternative hypothesis**', or H1, or the 'maintained' or 'research' hypothesis. This rejects the null hypothesis. A test only allows you evidence to support the rejection of the null hypothesis, to say with some confidence that the effect is unlikely to be a fluke.

Put another way, if the null hypothesis is rejected, then you can feel that there is some indirect evidence to support the alternative hypothesis. So, hypothesis testing does not directly support the effect under investigation; it merely attempts to disprove the null hypothesis, that your result is a fluke of some type. Saying that a result is 'significant' is something of a lay term in classical statistics; you might use it in applied research, but not under the eyes of your tutor! (Do note that the tests themselves are examining the null hypothesis, that there is no peculiar variance. Unlike you, the computer does not care about your cherished effect!)

The statistic that we most often read in classical statistics to see if the null hypothesis may be rejected is the **p value**. This is a decimal number between 0 and 1 which your test will generate. The smaller the number, the more likely it is that we would declare 'significance' (that the null hypothesis can be rejected). So, to give two more or less random examples, $p = .783$ is a large value whereby the results are almost certainly useless for experimental purposes; there is emphatic support for the null hypothesis. Nearer to the other end, $p = .007$ is really quite small (we've been waiting for you, Mr Bond) and, in most cases, we would feel justified in rejecting the null hypothesis.

Well that's all right then. We know that big p values mean that our effects are, well, ineffectual (the null hypothesis again) and that small numbers mean that we are famous.

But this raises a question or two. How big does a number have to be to mean that our effect is a fluke? And are there different levels of what we might call 'small'? And was I such a nuisance as to say that we would feel justified in rejecting the null hypothesis with p = .007 only 'in most cases'?

Enter the **critical value**. It is possible for the experimenter to pre-set a value at which the null hypothesis would be rejected. The typical critical value in social science experiments is p < .05 (p is smaller than .05) and it is quite likely that you will tend to use this in your course. So a p value of .046 would allow us to reject the null hypothesis (victory is ours), and a p value of .06 would not. This was suggested by the statistician RA Fisher as a useful rule of thumb, all other things being equal:

> ... it is convenient to draw the line at about the level at which we can say: "Either there is something in the treatment, or a coincidence has occurred such as does not occur more than once in twenty trials". Fisher (1926)

'Once in twenty' is of course the same as 5 in 100 (.05), but please do not treat these proportions as real by putting them into calculations or claiming that you have 95% likelihood or anything like that – they are only probabilities.

There are times when an experimenter would like to set a lower critical value. Commonly seen critical values are p < .01 and p < .001, although others are also possible. The lower the critical value, the smaller the p value required to support rejection of the null hypothesis. When it comes to aviation safety, for example, I would hope that a critical value as high as p < .05 would not be set for testing a mission-critical piece of equipment.

The critical value, particularly at the level p < .05, has come under considerable criticism, particularly in the area of psychology. It should be remembered that failure to find a small enough p value does not mean that an effect definitely does not exist. Also, the opposite is possible, that what appears to be an effect is in fact a fluke or the effects of other variables, statistical noise if you like. There is such a thing as the **Type 1 error**, believing that an effect exists when it doesn't; this is why you should be careful not to run too many tests within a study, as some results are likely to be flukes. **Type 2 errors** are the opposite, rejecting an effect which does in fact exist; this can be the effect of being too dogmatic about a cut-off. Either type of error is possible as a consequence of using the wrong test.

However, the continuing usage of the $p < .05$ rule of thumb over so many years does suggest a history of effectiveness in picking up effects (Bross 1971). Assuming that we accept $p < .05$ as a useful general guide, we are still left with questions such as 'do we reject a p value of .052' and 'is a value such as .045 always a meaningful effect?' Also, is a result significant but not of much use to the world? If you work with 'big data', you will find lots of very small p values and will wonder what to do with them all.

This brings us to another statistic, the **effect size**. This tells us how much of the variance in the sample is likely to be because of the effect. If, for example, you get an effect size such as .671, we can say that the effect is likely to be responsible for 67% of the variance in the sample. At other times, you may feel able to reject the null hypothesis, but find an effect size so small as not to be particularly useful in the real world.

Now, there is nothing wrong with reporting the test statistic, the p value and if possible the effect size and letting the reader decide if the value is acceptable. In fact, I would advocate it, except for the fact that tradition and custom, particularly among publishers of research, has made things much messier. How small a p value is small enough to reject the null hypothesis and report our experiment to the world? Does the publisher insist on neat tables, with p values accompanied by asterisks and critical values (* = $p < .05$, ** = $p < .01$, and so on)?

Regardless of your feelings about citing a critical value, you could publish the actual p value alongside it and indeed the effect size. Do not worry about these terms, as you will see various examples of NHST in action, which should accustom you to the concepts and practice. *

One-tailed and two-tailed hypotheses

I'm sorry, but there is another related concept that has an effect on how we consider our data. If you have good reason to know the direction of an effect before running the test – we are for example quite sure that the mean of A will be smaller than B (but not necessarily how much smaller) – then you may choose a one-tailed hypothesis. A good reason is a clearly explainable rationale or theoretical underpinning for the prediction. Otherwise, you should opt for a two-tailed hypothesis.

Two-Tailed Test

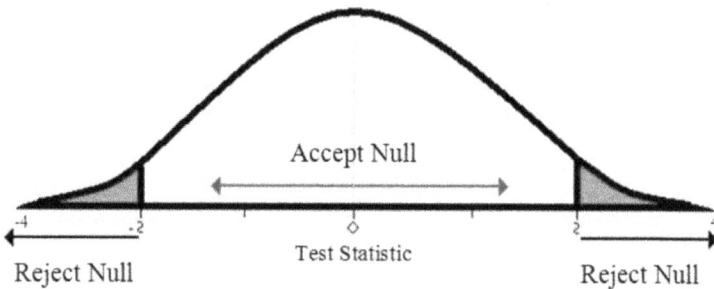

Accept Null

Test Statistic

Reject Null   Reject Null

As you will see from the chart, the result must be more rigorous for a two-tailed hypothesis. By allowing for potential variance on either side of the mean, we are having our cake and eating it. This also means a more rigorous approach to the p value, which will be bigger than if a one-tailed hypothesis had been chosen.

---

*Also sometimes demanded are confidence intervals, to be discussed later in the book.

Chapter 4 – Null hypothesis significance testing

One-Tailed Test

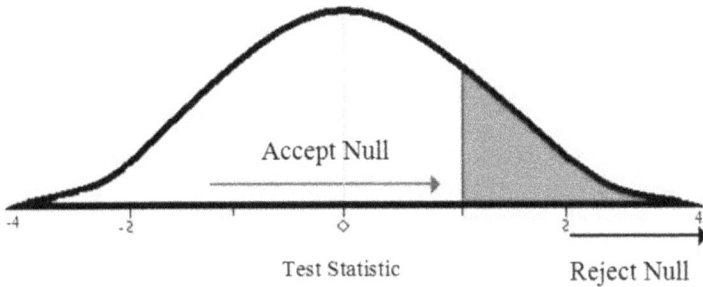

Less variance is under consideration when we are only considering one tail of the distribution curve. This being the case, a more lenient attitude is taken to the results. In practice, this means a reduction in the p value, so it is easier to reject the null hypothesis.

If this is used inappropriately, it is possible to end up with a Type 1 error, wrongly claiming significance. If in doubt, use a two-tailed hypothesis. One thing that you definitely should not do is to opt for a one-tailed hypothesis just to 'find significance'. This is frowned upon.

The next test is one that you probably won't use much. Its main purpose here is to allow you to accustom yourself to the software and to apply some of the concepts discussed.

One Sample T-Test – does a sample belong to a population?

This test is designed to test whether or not a sample belongs to a population when we know the mean of that population. In this case, I believe that salaries should have increased since the previous records and want to see if this is a reasonable claim. I know that the previously recorded mean was 22,000 and I want to find out if current incomes (in thousands) really are bigger.

Having opened the Basic.csv file with Jamovi, we press the T-Tests tab and select One Sample T-Test from the T-Tests tab.

I have transferred the Salary variable to the right-hand box, which is where the action takes place. Most importantly, I have entered 22 as my Test value (click a table on the display screen on the right in order to update the results). I have also gone for a one-tailed hypothesis, stating that the data is greater than the test value. I have also chosen Descriptives, Mean difference, Effect size and a Normality check (the Shapiro-Wilk test).

One Sample T-Test

		Statistic	df	p	Mean difference		Effect Size
salary	Student's t	2.23	9.00	0.026	7.90	Cohen's d	0.706

Note. H_a μ > 22

Normality Test (Shapiro-Wilk)

	W	p
salary	0.938	0.530

Note. A low p-value suggests a violation of the assumption of normality

Descriptives

	N	Mean	Median	SD	SE
salary	10	29.9	27.5	11.2	3.54

This is the full display of results on the right-hand side of the program. The T-Test result shows us a p value of .026. We have reason to reject the null hypothesis. The Mean Difference is for convenience: it shows us the difference between means of the test value I chose and the mean of the data. We have quite a large effect size (Cohen's d). Looking at the bottom, we get the descriptive data we are most likely to report, in particular the Mean for the data and the standard deviation.

In the center, we find the Shapiro-Wilk test. As the p value is a big one, it seems reasonable to assume a normal distribution. So Student's test, a parametric test, is fine to use.

If the Shapiro-Wilk test had been significant, with a very small p value, probably < .05, I would have chosen the 'Wilcoxon rank' option instead, as this is the non-parametric equivalent of the Student test.

Note that Shapiro-Wilk is not just for use with the t test. You can transfer other variables to the right to get normality tests for all of them.

The two-tailed hypothesis revisited

Before we leave this chapter, let us return to the two-tailed hypothesis. If I had no theoretical reason to believe that the new data set was different in a particular direction from the test value, I would have opted for the Hypothesis at the top, merely that the data is different from the test value. This might have been because I was uncertain, or perhaps that I had expected the salaries to be approximately the same and wanted to test this; a non-significant result would indicate no substantive change. If you switch to the two-tailed hypothesis in our worked example, you will see that the p value rises to .052, not quite within the suggested limits offered by Fisher. *

---

*In such a case, you could see what the Bayes factor option says about the results. To interpret this, use the Bayesian reporting table in Chapter 17.

Part 2

Statistical testing fundamentals

Chapter 5 – Tests of differences

Design considerations for the analysis of differences

We need to consider research design every time we think about posing a question. This also applies to choosing the most appropriate test. Of particular concern is the issue of Related versus Unrelated design. You will come across other terms in Jamovi and the statistical literature:

Related design	Unrelated design
Same subjects/participants	Different subjects/participants
Paired design	Independent design
Within–subjects design	Between–subjects design
Paired	Unpaired
Matched	Unmatched
Repeated measures (the same over time)	
Panel data (the same people over time, but a term used in business statistics / econometrics)	

As an example, let us suppose that we were planning a study of attitudes to physical education in a region's schools. Before the study begins in earnest, we might want to find out if pupils' energy levels tend to be different at the start and the end of the school week.

We may choose a **related design**, studying the same individuals. We could have all pupils tested on Mondays and Fridays, as perhaps energy is affected by the experiencing of a school week. A related design is generally preferred because it eradicates individual differences, but it may not always be feasible, if we need individuals to be fundamentally different, as in studies of gender or status differences, or perhaps it is impractical in the particular context. In this example, we run the risk of people becoming bored with the tests or some form of experimental contamination in the intervening period. With two conditions, we would consider using the paired samples *t* test or the non-parametric Wilcoxon signed rank test.

In an **unrelated design**, in which we study different people, we may decide to have half of the schools tested on a Monday and the others on a Friday. Testing different people rids us of carry-over effects. However, such a design introduces the problem of individual differences, in this case applying to both the students and potentially different school environments. At this stage, for two conditions, we would consider using either the independent samples *t* test or its non-parametric equivalent, the Mann-Whitney *U* test. These tests of independent samples to some extent allow for individual differences.

Some researchers like to attempt **pairing**, also known as **matching**. Although the individual participants are not the same, they are grouped for characteristics which are deemed relevant to the study. In this example, we may decide to pair up schools from similar social milieu, and maybe students of the same gender and body type. In such a situation, this could be seen as a related design, so the use of tests for same subjects (paired samples *t* or Wilcoxon) would be deemed appropriate.

In addition to the question of related and unrelated design, we also need to consider if we are using *more than two conditions*. So far we have considered only two conditions. Let us say that we decide to run our tests on a Wednesday as well as on Monday and Friday, giving 3 conditions, or perhaps every weekday, making 5 conditions. Another example would be if we were to try out three different interventions to improve engagement in exercise (Intervention 1, Intervention 2, Intervention 3), or two interventions and a control condition. For more than two conditions, we would consider using, for the related design, either the Repeated Measures ANOVA or the non-parametric Friedman test; for the unrelated design, we would consider ANOVA or the non-parametric Kruskal-Wallis test.

Yet another consideration, which helps us to narrow down to the test of choice, is *the type of data* being examined. In this chapter, data must at least be comparable numerically (categorical, or 'nominal' data appears in Chapter 7, which covers frequencies of observation). If the data is continuous, it is a candidate for parametric tests, such as the *t* test and ANOVA (Analysis of Variance), as mentioned previously, but we also need to consider the assumptions of normality of distribution and homogeneity of variance covered in Chapter 3. If these assumptions are not met, or the data is ordinal, then we should abandon parametric tests and should consider non-parametric tests, including the Wilcoxon, Mann-Whitney, Friedman and Kruskal-Wallis.

Some research design terminology applied

When we deliberately set up an intervention, including the allocation of cases to different conditions, we are running an experiment (with or without a laboratory). We manipulate variables, things which are changeable (variable). We actively manipulate an independent variable, the effect of the time of the week in the recent example; it could be, however, the type of intervention, or a single intervention over time. We observe the effect of manipulating the independent variable by looking at changes in the dependent variable, the measure. The dependent variable could be, for example, test results, physiological measures and speeds.

The independent variable is varied by the experimenter. The dependent variable, the measure, is data which is dependent on such variations. The terms independent variable and dependent variable should, strictly speaking, only be used when referring to experiments, but in practice they are used much more widely.

Very many 'real world' analyses of differences are quasi–experimental. We do not manipulate variables ourselves, but use records or observations without allocating groups in any organized way. If we ran a quasi–experimental version of our study, the effect of which part of the week would be called a predictor (instead of independent variable) and the test result would be the criterion (instead of dependent variable). The criteria for measurement are likely to include the number of students attending classes, ratings, social interactions, reduction in accidents and so on.

While the terms predictor and criterion should be used in non––experimental research, they are often interchangeable with independent and dependent variables in the literature. Both terms are used in the following example (ignore the results; there are too few observations).

	Independent Variable / Predictor: Part of the week	
	Condition 1: Monday	**Condition** 2: Friday
	40	50
Dependent Variable /	30	40
Criterion:	45	45
Attitude scale	60	50
	45	43

To practice using the statistical tests in this chapter, we use the variables held in the file Differences.csv, although the data is also presented on the page so that you can create your own files should you wish.

Tests for same subjects

Paired Samples *t* test: a parametric test for two conditions, same subjects

Rugby referees are at different times asked to take two very similar tests of judgement, both validated as of equal difficulty. Each test consists of various video examples of difficult scenarios, shown one after the other at high speed. We are interested in the effects of alcohol on judgement. On the first week, before seeing one of the tests, half of the referees had consumed two units of alcohol at a set time before the showing, while half had consumed four units. The following week, they saw the other test with alcohol dosages reversed.

All received scores for their performances on the tests. If there was a noticeable difference between the scores, we want to know if such a difference was statistically significant. As each individual underwent both conditions at some point, this is a related design.

| | | Predictor: Alcohol consumption | |
	Experimental participants	Condition 1: Alcohol – 4 units	Condition 2: Alcohol – 2 units
Criterion:	1	52	60
	2	53	34
	3	47	38
	4	40	52
Recall score (as a percentage)	5	48	54
	6	45	55
	7	52	36
	8	47	48
	9	51	44
	10	38	56

Open Jamovi. Opt for the File tab and use the Browse option to find the Differences.csv file.

Alcohol4	Alcohol2	AlcoholNil
52	60	62
53	34	46
47	38	47
40	52	39
48	54	58
45	55	56
52	36	68
47	48	56
51	44	67
38	56	60

The two relevant variables are Alcohol4 and Alcohol2. The next one, AlcoholNil, will extend our example to three variables when we use the Repeated Measures ANOVA a little later.

Press the T-Test tab at the top and choose Paired Samples T-Test from the menu.

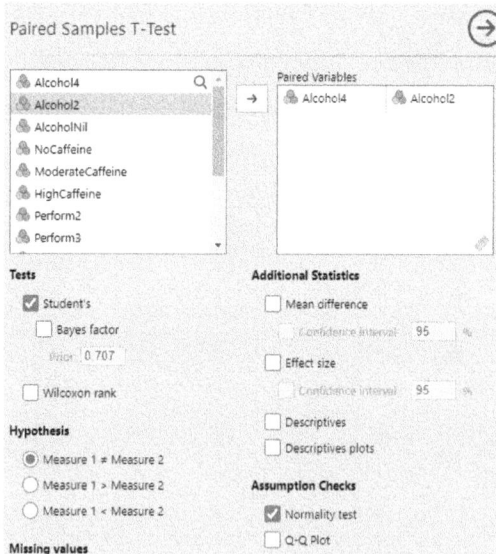

Alcohol4 and Alcohol2 have been shifted from the list of variables on the left to the box on the right. To do this, you can use the transfer (arrow)

button, or double-click, or drag and drop. We retain the default option, Student's *t* test. [*] The Normality option examines the distribution of data using the Shapiro-Wilk test.

Paired Samples T-Test

			statistic	df	p
Alcohol4	Alcohol2	Student's t	−0.101	9.00	0.922

Test of Normality (Shapiro-Wilk)

			W	p
Alcohol4	-	Alcohol2	0.945	0.606

Note. A low p-value suggests a violation of the assumption of normality

If Shapiro-Wilk is insignificant, as it is here, then a normal distribution is assumed and the *t* test can be considered valid. The *t* test result has a huge *p* value, 0.922, so there is clearly no significant difference between the variables (if you use the Descriptives option, you will see means of 47.3 and 47.7 respectively). Using the correct terminology, we do not have evidence to support rejection of the null hypothesis.

We conclude here that the scores did not differ significantly. (Reporting note: you would refer to the 'statistic' as *t*.)

The Wilcoxon test: a non-parametric test for two conditions, same subjects

Female synchronous swimmers are rated in an experiment involving the use of caffeine. On one occasion, they take no caffeine prior to a session; on another they consume a moderate amount. Performance

---

[*]Sorry, this was not designed especially for students; it was named after a chemist at the Guinness brewery who used Student as a pen-name.

is measured on an agreed 5-point rating scale. Higher ratings mean positive outcomes overall; lower ratings are negative.

The scale did not undergo preliminary calibration; for a Likert scale to be considered as continuous data, we need to believe that it is proportional, for example that the value of a 4 rating really does approximate to twice as much as a 2. Without calibration, a non-parametric test is preferred. Given a small data set, we settle for a critical value of $p < .05$; as we are not really sure which service would be more effective, we opt for the more rigorous two-tailed hypothesis.

		Predictor: levels of caffeine consumption	
	swimmer	Condition 1: No caffeine	Condition 2: Moderate caffeine
Criterion:	1	4	5
	2	3	3
performer	3	2	4
(1 to 5 Likert scale)	4	4	5
	5	3	5
	6	4	2
	7	3	3
	8	5	4
	9	3	5
	10	4	5
	11	3	5
	12	2	4
	13	2	5

We use the T-Test tab and again choose the Paired Samples T-Test option. As in the previous example, transfer the relevant variables to the right, here the NoCaffeine and Moderate variables. Under the Tests options, select 'Wilcoxon rank', the non-parametric test we require, deselecting or ignoring the Student's test, which is the parametric test we used previously. From the Additional Statistics section, choose Effect size; also select Descriptives.

Paired Samples T-Test

			Statistic	p		Effect Size
NoCaffeine	ModerateCaffeine	Wilcoxon W	10.0 [a]	0.041	Rank biserial correlation	−0.697

[a] 2 pair(s) of values were tied

Descriptives

	N	Mean	Median	SD	SE
NoCaffeine	13	3.23	3	0.927	0.257
ModerateCaffeine	13	4.23	5	1.013	0.281

We are interested in the Wilcoxon's W statistic (ignore the T-test header). First, we look at the p value: .041 comes within the $p < .05$ critical value.

If we had expected beforehand, theoretically and without eyeballing the data, that the use of caffeine was going to achieve a higher rating than non-usage, we could have selected a one-tailed test. This would have been achieved by changing the Hypothesis to Measure 1 < Measure 2. The resulting p value for the Wilcoxon test, .02, would be smaller.

A thinking point: If you look at the t test result for this pairing, you will find that the Wilcoxon is more conservative. This is intentional, given the roughness of the data. However, if you use data suitable for parametric tests, you will often find that both the Wilcoxon and t tests give the same results. I think this important when considering the claims of those who advocate the use of parametric tests in all circumstances. They are 'more powerful', we are told, and non-parametric tests are old-fashioned, always a reason for throwing away something that works!

As we have evidence to support rejection of the null hypothesis, we are interested in the effect size (the option is to be found under 'Additional Statistics'). Unlike 'significance', this measures the magnitude of the difference between the variables. Effect sizes are to some extent dependent on context, but generally for t tests, from 0.2 to 0.3 is considered small, 0.8 upwards is large (this statistic can exceed 1.0); in-between values are medium, as in this example. Ignore the minus in this example; if we had transferred the variables the opposite way, this would become a positive (0.697). In either case, this would be deemed a medium-sized effect.

Jamovi offers you a set of reportable descriptive statistics. As you already know, the median is the correct statistic for non-parametric tests. However, you will find when handling Likert scales that the mean gives superior differentiation when handling results that are less clear-cut than those in our current example.

Another facility that you can try out is the 'Mean difference' option within 'Additional Statistics', another way of measuring the effect. For parametric tests, this is the difference between the means, subtracting one mean from the other. For non-parametric tests, Jamovi uses the pseudomedian. [*]

Repeated Measures one-way ANOVA: a parametric test for more than two conditions, same subjects (also known as the Within-subjects one-way ANOVA)

As the title 'repeated measures' suggests, this type of test can also be used for the analysis of time series; do note, however, that there is quite a lot to know about time series and that some reading around the subject is advisable. In this example, I will be looking at ANOVA for different conditions; although there is a time difference here, it is secondary to a qualitative difference between conditions.

Here, we use our earlier t test example, where we contrasted the results of tests of judgement with differing levels of alcohol consumption. Here we decide to include a control condition when the referees judged yet another parallel test without any alcohol (as with the previous example, the results are fictitious).

Predictor: Alcohol levels or no alcohol

---

[*]The pseudomedian (the Hodges-Lehman estimator to its friends) comes nearer to the mean difference than would a straightforward difference between medians. If you calculated the median difference, it would in this case be 2; if you used the t test the mean difference would be 1. Here, the Wilcoxon pseudomedian statistic is 1.5.

Condition 1: 4 units **Condition 2:** 2 units **Condition 3:** No alcohol

The additional results are: 62, 46, 47, 39, 58, 56, 68, 56, 67, 60, which are to be found in the variable AlcoholNil from the Differences.csv data file.

First we look at the descriptive data. *You should always do EDA (exploratory data analysis) before running an ANOVA.* We need to check the assumptions of normal distribution and that the data is suitably measurable.

For the Shapiro-Wilk test of normality, you can use the Normality option within Paired T-Tests and try it out on the different pairings of variables. In this particular example, you will find that all of these are non-significant. Another option is to go to the Descriptives tab, opt for Descriptive Statistics and open Statistics, using the Distribution statistics for all three variables to check Skewness and Kurtosis. While you are there, you can also check that your data is legitimate by looking at the Dispersion values; these will help you to check for peculiar or illegitimate data.

However, for ANOVA, there are more assumptions to meet. To see if the data meets these assumptions, we need to set up the ANOVA test itself. Press the ANOVA tab and select Repeated Measures ANOVA from the menu. You should see this mysterious-looking interface:

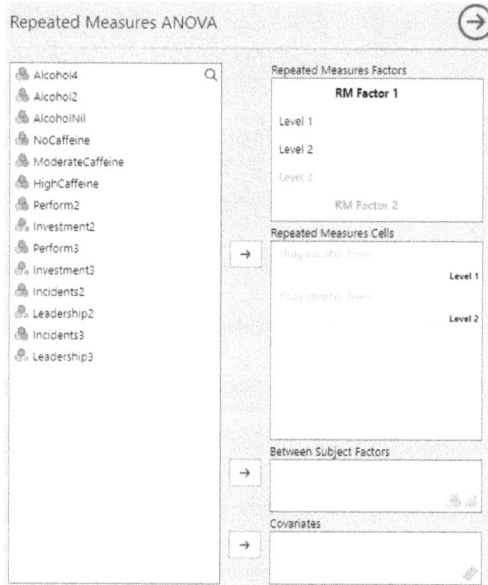

This is a little more difficult to set up than the *t* test, but compared to doing it with SPSS, it's a breeze.

The first thing we should do is to replace RM Factor 1 with a more meaningful name; using the title of the predictor / independent variable would make sense. In the one-way ANOVA, this will assist your reader's understanding of the output; when you get to use a two-way (or three-way) ANOVA, you will need it to assist your own understanding!

Then replace the levels (another term for conditions) starting with Level 1, by typing in their titles. If you use post hoc tests, or the descriptive plot, these will appear. You will probably want to provide somewhat more detailed names than those given in the file, but if they are too long, they may look confusing on some of the output.

You only need RM Factor 1 for one-way ANOVA; further factors come into play when you use a factorial ANOVA (to be covered in Chapter 10). Similarly, you only need as many levels as the conditions you are analyzing (which reminds me: if you use too many variables, you are increasing the danger of a Type 1 error, thinking that you have a significant finding when you don't).

At this stage, the top left of the Repeated Measures ANOVA interface should look like this:

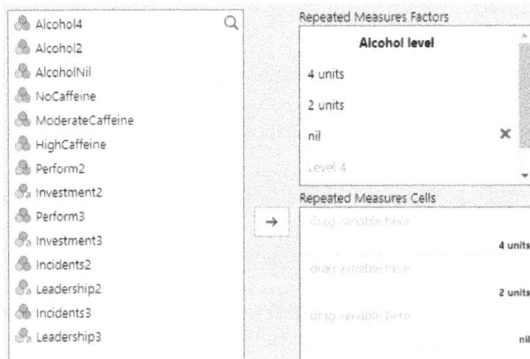

The titles of the levels typed in underneath the factor name have automatically been transferred to the right-hand side of the Repeated Measures Cells box. Then comes the easy bit: transferring the variables themselves to the left-hand cells. Do be careful as your data sets get more complicated: the transferred variables on the left must match the levels on the right.

The variables are transferred by using the arrow or dragging. So you should then have this:

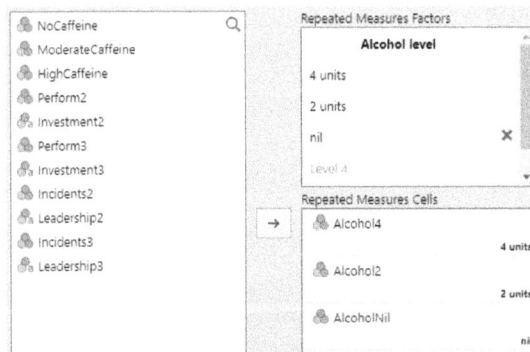

You will notice that I have not shown the lower boxes. The Between Subject Factors box only comes into play when we use a mixed design ANOVA, combining within– and between–subjects factors (Chapter 10). The Covariates box is for variables of no particular interest but which might influence interactions.

Although statistics immediately appear on the ANOVA table on the display on the right, you need to ignore them for the moment. The results of an ANOVA are likely to be misleading if the data does not meet certain assumptions. As well as the usual assumptions for parametric tests in general, the repeated measures ANOVA has an assumption of sphericity (equality of variances between each pair of levels). Problems with sphericity do not mean that you won't be able to use the ANOVA, but you would need to use a non-standard read-out. So before looking at the results, you need to open the Assumption Checks section and select Sphericity tests. At the moment, we are only interested in Mauchly's W. If this is significant at the level of $p < .05$, then the assumption is violated.

Test of Sphericity

	Mauchly's W	Approx. X²	df	p	Greenhouse-Geisser ε	Huynh-Feldt ε
Alcohol level	0.846	1.342	2	0.511	0.866	1.000

In this case, however, a p value of .511 is healthily non–significant, so we can ignore sphericity as a problem. The figures to the right are only of relevance if p is significant. If you do find yourself with a sphericity problem, refer to Chapter 10, where the problem arises in the ANOVA mixed design worked example.

Before considering the results of the ANOVA itself, let us quickly consider the concepts which are central to analysis of variance. The variance is the variability of data around the central tendency (usually the mean); if observations tend to vary a lot from the mean, the variance is large, and vice-versa. Analysis of variance calculates how much variance comes from independent variables and how much is due to error (error variance). The calculation, the variance divided by the error, is the F ratio, referred to in the output as 'F'. Essentially, the bigger the F ratio, the more likely it is that the effect is a significant one.

The ANOVA table shows an F ratio of 3.64 and a p value of 0.047. Assuming that we accept a critical value of $p < .05$, our effect may be considered to be a significant one. There is evidence to support the rejection of the null hypothesis of no mean differences across the three conditions.

We can also look at the effect size. One rule of thumb for ANOVA effect size is that 0.01 is small, 0.06 is medium and 0.14 is large. Eta squared indicates a large effect of .208. Partial Eta squared is preferred for factorial ANOVA.

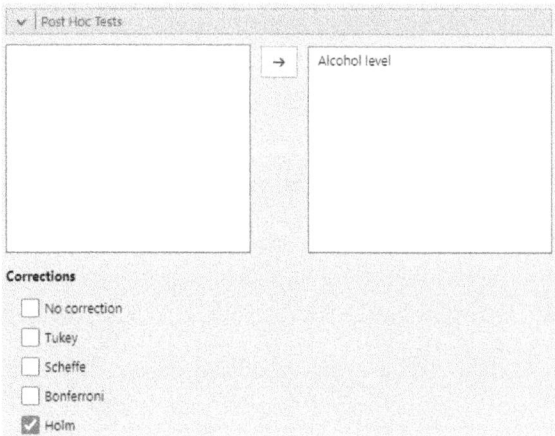

If you wish to look at individual pairings, open the Post Hoc Tests section and transfer the relevant variable to the right. The Bonferroni correction is often the default in statistical literature, but it is nowadays considered too severe.

Post Hoc Comparisons - Alcohol level

Comparison							
Alcohol level		Alcohol level	Mean Difference	SE	df	t	P_{holm}
4 units	-	2 units	−0.400	3.96	9.00	−0.101	0.922
	-	nil	−8.600	2.81	9.00	−3.065	0.040
2 units	-	nil	−8.200	3.91	9.00	−2.097	0.131

There is reason to believe that one of the pairings is of particular importance. These tests will be discussed a little more towards the end of the chapter when applied to between-subjects ANOVA and in far more detail towards the end of Chapter 10.

Friedman: a non-parametric test for more than two conditions, same subjects

If the data fails to meet the assumptions of continuous data and normality, use the Friedman test instead of Repeated Measures ANOVA.

In addition to the two conditions examined by the Wilcoxon test, swimming without caffeine and with moderate caffeine, a further condition involves the consumption of a relatively high level of caffeine. As the scale has not been calibrated, a non–parametric test is preferred.

Should you wish to input your own data, the new condition contains the following data: 5 3 2 2 2   2 1 3 2 4   2 3 1

Press the ANOVA tab and select Friedman from the menu.

Repeated Measures ANOVA (Non-parametric) →

	Measures
Alcohol4	NoCaffeine
Alcohol2	ModerateCaffeine
AlcoholNil	HighCaffeine
Perform2	
Perform3	
Incidents2	
Incidents3	
Investment2	

☑ Pairwise comparisons (Durbin-Conover)

☐ Descriptives

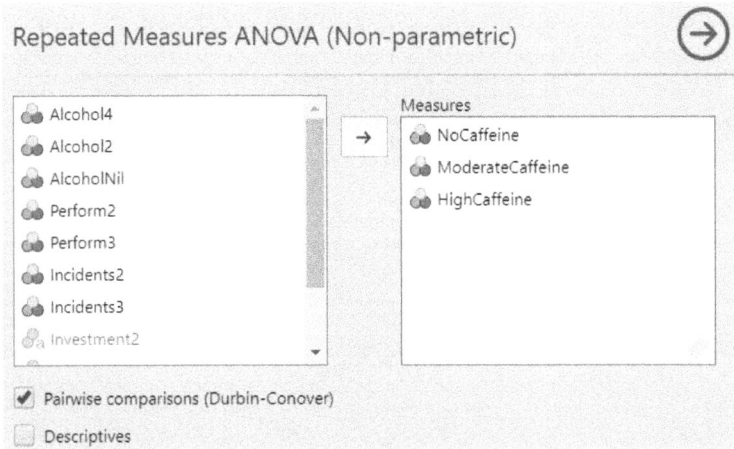

The relevant variables are transferred to the right-hand box and off we go.

Friedman

χ^2	df	p
12.7	2	0.002

Pairwise Comparisons (Durbin-Conover)

			Statistic	p
NoCaffeine	-	ModerateCaffeine	2.61	0.015
NoCaffeine	-	HighCaffeine	2.17	0.040
ModerateCaffeine	-	HighCaffeine	4.78	< .001

The Friedman test provides evidence to support the rejection of the null hypothesis. A particularly low p value pertains to the Moderate and High pairing. The direction of the effect is very clear when you examine the Descriptives plot. The medians for these variables are 5 for Moderate, 2 for High and 3 for NoCaffeine. Strictly speaking, the median is the better statistic for non-continuous data; however, should the medians be close enough together to be indistinguishable, I would be less of a purist and use the mean as the measure of central tendency.

The multiple comparison tests provide evidence to support the rejection of the null hypotheses for each of these pairings. A particularly strong difference is indicated for the Moderate and High pairing; $p <$ .001, not surprising given their respective median values. [*]

Tests for different subjects

Independent Samples T-Test: a parametric test for two conditions, different subjects

We are interested in the difference between high and low sporting investment countries as reflected by measures of performance of young table tennis players. A measure has been created using a composite of different skill tests; higher scores represent better performance levels. The independent variable is the level of investment, high or low.

---

[*]For reference purposes, the test used here produces the same results as StatsDirect's (2011) application of Conover (1999).

Table tennis players	Criterion / dependent variable	Predictor / grouping
	Score on test Name: **Perform2**	Investment level Name: **Investment2**
1	80	high
2	68	high
3	77	high
4	78	high
5	85	high
6	82	high
7	79	high
8	76	high
9	77	high
10	83	high
11	84	high
12	82	high
13	81	high
14	80	high
15	56	low
16	69	low
17	73	low
18	70	low
19	61	low
20	65	low
21	59	low
22	60	low
23	53	low
24	61	low
25	62	low
26	71	low

In the previous worked examples, we performed data entry for same-subjects design, which was quite simple. However, as previously noted in Chapter 1, data entry for between-subjects design is a little less straightforward: as an observation or participant needs its own row, different conditions need a demarcation method. Each case has to have its own grouping variable, whether using names (such as medium) or numbers. So the first 14 are given the 'high' level, while 15 to 26 are 'low'.

In the Differences.csv file, as we have a few simple projects together, only the active variables are included, here the numeric variable Perform2 and the grouping variable Investment2. In real life research, you would save identification variables (here, 'Player' or 'Team'). When you are involved in a complex piece of research, case numbers are particularly helpful in error-checking; when you are rushed and/or tired, errors do creep in.

Notice that in this study, there are different numbers of cases in the different conditions, 14 'high' investment and 12 'low'. Only same subject studies are required, quite logically, to have the same numbers in each condition.

To run the test, press the T-Test tab and select Independent Samples T-Test from the menu.

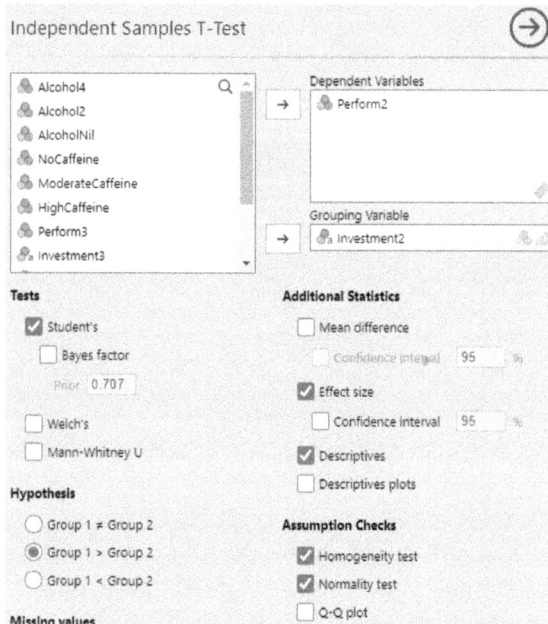

The Investment2 variable has two conditions, high and low, and is named to distinguish it from another variable to be used later which has three conditions. It has been transferred to the Grouping Variable box. The corresponding Perform2 goes into the Dependent Variables box.

You will find that you have different preferences as you get used to the software. Let us consider the choices taken above.

I would always use the Assumption Checks, Normality and Equality of variances (also known as homoscedasticity), as these will determine which test should be used. The default *t* test, Student's, is a parametric test; the data should meet both assumptions as well as being continuous. If the data is clearly not continuous, or the normality assumption has not been met, then you definitely want Mann-Whitney U as a non-parametric test. If the data is continuous and generally normally distributed, but the equality of variances assumption has not been met, or the numbers of observations in each group differ greatly, then Welch's test is to be preferred as an intermediate test between Student's t test and Mann-Whitney.

Independent Samples T-Test

		Statistic	df	p		Effect Size
Perform2	Student's t	7.73	24.0	< .001	Cohen's d	3.04

Note. Ha μ high > μ low

Assumptions

Normality Test (Shapiro-Wilk)

	W	p
Perform2	0.976	0.787

Note. A low p-value suggests a violation of the assumption of normality

Homogeneity of Variances Test (Levene's)

	F	df	df2	p
Perform2	3.35	1	24	0.080

Note. A low p-value suggests a violation of the assumption of equal variances

Looking first at the Assumption Checks, we can see that both tests have p values greater or equal to .05, so it is reasonable to stick with the default Student's t test for our two variables. Note that I have chosen a one-tailed hypothesis (Group 1 > Group 2, based on alphabetical order) as the direction of the effect was very much expected. The t

test indicates that we can reject the null hypothesis. (The Welch and Mann-Whitney options show similar results in this example.)

The Additional Statistics are worthy of some study, in that they tell us more about the data. In particular, it is worth examining Effect size. The output shows the Cohen's d statistic. For t tests, Cohen's d from 0.2 to 0.3 is considered small, 0.8 upwards is large (this statistic can exceed 1.0); in-between values are medium. The current effect size of 3.04 is huge. The Descriptives are well worth having as well: apart from reminding you of which variable is larger, you are likely to want to cite the means and standard deviations (SD) in your reports.

Options that you may also consider include 'Mean difference' and 'Confidence interval'. The first is a measurement of difference between the two samples, by subtracting one mean from the other; like the mean and the median, this is a 'point estimate', a single statistic which acts as a representative value but has some limitations. The confidence interval is intended to show a range of likely values around the point estimate, citing two figures, the lower and upper confidence limits. Note that if your Hypothesis setting is one-tailed (in this case, Group 1 > Group 2) then one limit will be infinity 'Inf'); to ascertain both confidence limits, change to a two-tailed hypothesis.

Perform2

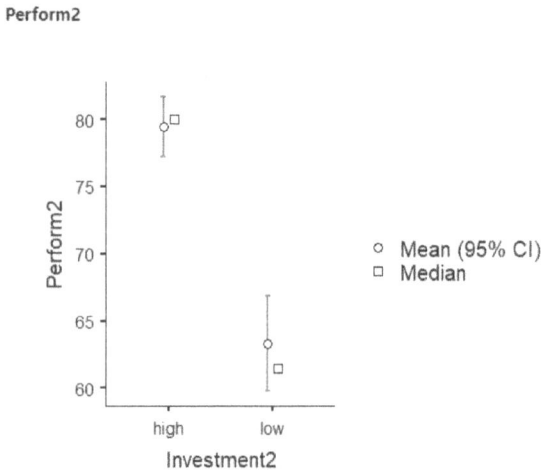

The chart created by the Descriptives Plots option shows the clear disparity between the two conditions. The bars represent the confidence intervals, showing a range of likely values. Do note, however, that the confidence intervals do not show the whole range of possible values - note the 95% cited in the chart – and they are not without their critics (Morey *et al* 2016).

The Mann-Whitney test: a non-parametric test for two conditions, different subjects

We are interested in the prevalence of fighting within men's ice hockey matches and wish to know if this is affected by leadership styles, adopting Lewin's model (Lewin *et al* 1939). Within each of several teams, the number of fights involving the players are counted over a short period of time. It seems possible that different types of leadership may affect individuals' self-control. (As usual, this is a fictional study.)

Criterion:	Predictor: Leadership style			
	Autocratic		Democratic	
Number of fighting incidents per team	Team 1	5	Team 11	6
	Team 2	4	Team 12	15
	Team 3	16	Team 13	4
	Team 4	6	Team 14	4
	Team 5	7	Team 15	6
	Team 6	22	Team 16	7
	Team 7	8	Team 17	16
	Team 8	9	Team 18	7
	Team 9	9	Team 19	5
	Team 10	8	Team 20	4

We convert the data, as necessary for a between-subjects design. These are the necessary variables (omitting the case numbers you would want in real life).

Leadership2	Incidents2
Autocratic	5
Autocratic	4
Autocratic	16
Autocratic	6
Autocratic	7
Autocratic	22
Autocratic	8
Autocratic	9
Autocratic	9
Autocratic	8
Democratic	6
Democratic	15
Democratic	4
Democratic	4
Democratic	6
Democratic	7
Democratic	16
Democratic	7
Democratic	5
Democratic	4

To run the test, press the T-Test tab and select Independent Samples T-Test from the menu. Transfer the relevant variables to the right. Incidents2 goes into the Dependent Variables box, with Leadership2 as the Grouping Variable.

As the test for normality of the distribution is significant, indicating a clear violation of the assumption, we require a non-parametric test, so under the Tests option, choose Mann-Whitney. (Choose Welch's t test if you have normality, but Levene's test is significant or there is a big difference between the numbers in each group.)

As there is no expected direction, we use the default two-tailed hypothesis. Choose Descriptives, as descriptive statistics are usually useful.

Independent Samples T-Test

		statistic	p
Incidents2	Mann-Whitney U	31.5	0.170

Group Descriptives

	Group	N	Mean	Median	SD	SE
Incidents2	autocratic	10	9.40	8.00	5.50	1.74
	democratic	10	7.40	6.00	4.43	1.40

A p value of 0.17 resulting from the Mann-Whitney means that we cannot reject the null hypothesis, that the two groups are not particularly different.

The descriptives table shows a difference in means. You might be tempted to look at the data derived from a one-tailed hypothesis, Group 1 > Group 2. If you did, you would see p = .085 for the Mann-Whitney test, which could be seen as a trend towards significance. However, unless you have a theory or rationale supporting a one-tailed hypothesis, the result really should not be considered as a demonstrable effect.

From the data we have, any difference between the leadership styles is unlikely to have a meaningful effect in this context.

Between-subjects one-way ANOVA: a parametric test for more than two conditions, different subjects

Let us extend our data from the Independent Samples T-Test example. The study previously examined the different performance scores for young table tennis players from high and low investment countries. We now think that perhaps we were not defining categories finely enough, so we examine a further sub-sample, players from countries with middle-range investment in sport.

New variables Perform3 and Investment3 have an additional 16 cases (27 to 42).

Table tennis player	Criterion / dependent variable	Predictor / grouping
	Score on test Name: **Perform3**	Funding regime Name: **Investment3**
25	62	low
26	71	low
27	70	middle
28	70	middle
29	73	middle
30	80	middle
31	81	middle
32	75	middle
33	75	middle
34	73	middle
35	81	middle
36	76	middle
37	75	middle
38	75	middle
39	73	middle
40	71	middle
41	72	middle
42	67	middle

Press the ANOVA tab and select ANOVA from the menu.

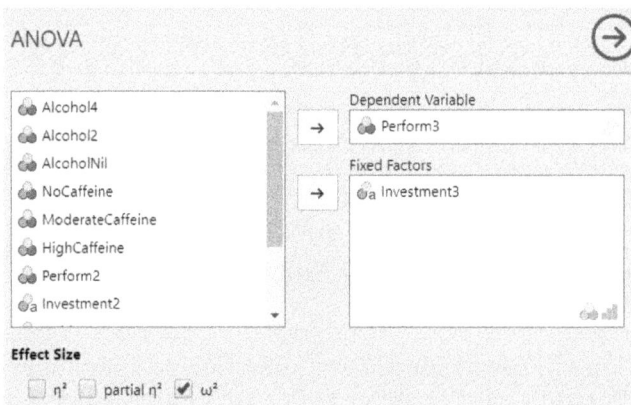

Transfer Perform3 to the Dependent Variable box and Investment3 to the Fixed Factors box. Assuming that you have already checked that the data is continuous and of a normal distribution, you examine homogeneity by opening Assumption Checks. You will also notice some squiggly items just below the variables boxes: these are effect size measures eta squared, partial eta squared and, selected, omega squared.

ANOVA

	Sum of Squares	df	Mean Square	F	p	ω^2
Investment3	1721	2	860.7	36.9	< .001	0.631
Residuals	911	39	23.3			

[4]

Assumption Checks

Test for Homogeneity of Variances (Levene's)

F	df1	df2	p
2.58	2	39	0.089

As Levene's test is not significant, homogeneity (also known as homoscedasticity) is not a problem. If the Levene test were to be significant, with a p value smaller than .05, then you could use the One-Way ANOVA option within the ANOVA menu and use Welch's ANOVA, which does not assume equality of variances. [*] If, however, the normality assumption is not met, then the non-parametric Kruskal-Wallis test should be used (introduced shortly).

Anyway, as we can continue, we see that the overall differences are significant. We have a large F ratio of 36.9 and p < .001. The larger the value of F, the easier it is to reject the null hypothesis. Residuals represent error variance.

Omega squared shows us an effect size of .631, 63% of the explained variance, so we have a very large effect size. A rule of thumb for ANOVA effect size is that 0.01 is small, 0.06 is medium and 0.14 is large.

The results for eta squared and partial eta squared are the same in one-way ANOVA but differ with multiple ANOVA. Here, if you look,

[*]Why not use One-Way ANOVA option, Fisher's variant? ANOVA does the same thing for one-way as for factorial ANOVA, so I only need One-Way for Welch; also the ANOVA version provides an effect size, eta squared.

they give the figure .654. Although partial eta squared is a favorite in many textbooks, there is now evidence which indicates that it over-estimates the effect size in smaller samples, the bias being quite robust even in samples of 100 cases (Okada 2013). So my guess is that omega squared is more accurate here.

If you then go down to the Additional Options section, you can opt for Descriptive statistics, which as well as showing the means and standard deviations, includes the number of cases in each group, which in complex projects will tell you if you have set up your data correctly.

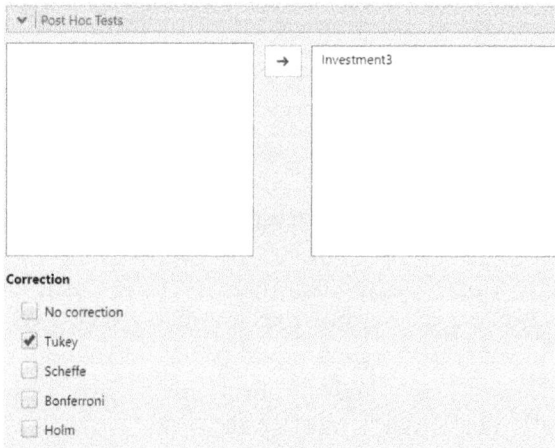

To see if these differences may be considered significant, Open Post Hoc tests. By transferring the grouping factor to the right, you can see the p value readings for pairings of levels.

Post Hoc Comparisons - Investment3

Comparison						
Investment3	Investment3	Mean Difference	SE	df	t	p_{tukey}
high	- low	16.10	1.90	39.0	8.47	< .001
	- middle	5.24	1.77	39.0	2.96	0.014
low	- middle	−10.85	1.85	39.0	−5.88	< .001

The larger the value of t, the easier it is to reject the null hypothesis. We generally don't just run a series of t tests, because of the possibility of fluke results creating false positives (Type 1 error), so an adjustment is often made in the case of multiple comparisons of pairs.

The default test in Jamovi is the Tukey test (a discussion of the choice of tests can be found in Chapter 10).

Open Estimated Marginal Means, transferring the grouping factor to the right.

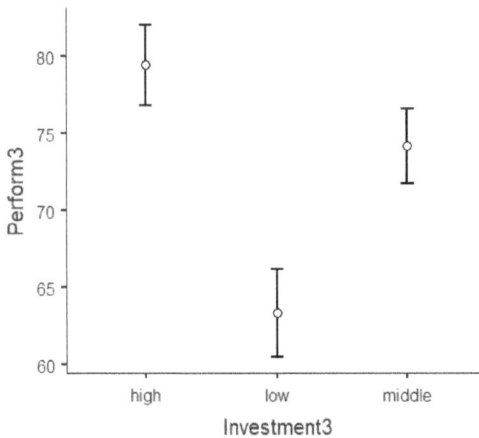

From the Tukey statistics and from the chart, you can see that the greatest effects are those contrasting with the 'low' condition.

According to this (fictional) data, those players from countries with low investment in sport appear in general to perform less well than those from countries with higher levels of investment. Those from high investment countries appear to perform better those from either of the other investment levels. The intermediary position of the players from middling investment countries does suggest that this finer gradation is worthwhile, although we cannot be certain from this study that there are no complicating factors.

Kruskal-Wallis: a non-parametric test for more than two conditions, different subjects

If the data fails to meet the assumptions for ANOVA, continuous data and normality, use the Kruskal-Wallis test.

In addition to the two conditions examined by the Mann-Whitney test, autocratic and democratic leadership styles, we now consider teams with a laissez-faire leadership. If you want to type in your own figures rather than use Differences.csv, the data for the laissez-faire condition are, 8 4 3 12 4   4 9 8 32 6

Press the ANOVA tab and select the Kruskal-Wallis test from the menu.

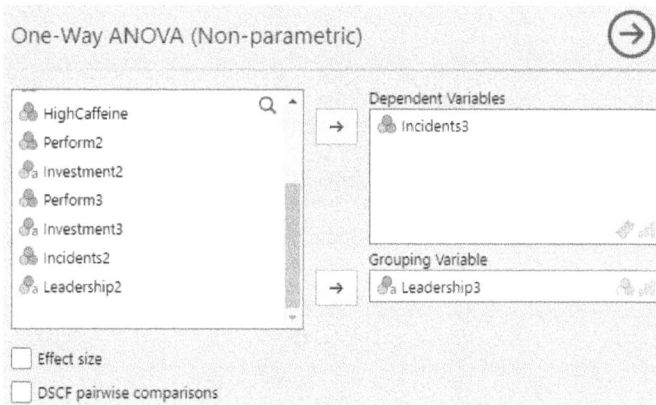

The p value is about 0.41; clearly the null hypothesis cannot be rejected. As I do not wish to dredge for data, we will not follow up with multiple comparisons (the DSCF option). Hsu (1996) notes that it is possible for pairings to be significant even when the overall test is not. However, I would like to adopt a considered approach to post hoc tests rather than their mere availability; see the discussion on multiple comparisons in Chapter 10.

Let us turn to a data set where we know that there are significant results.

The study used for the ANOVA examined young table tennis players' performances relative to different levels of investment funding regimes. We need to set up the test as before, but this time use Perform3 as the dependent variable and Investment3 as the grouping variable.

The result shows a p value of less than .001; we can also select the option 'DSCF Pairwise Comparisons'. This is the Dwass, Steel, Critchlow and Fligner test (let's call it the Dwass test for the sake of our sanity). This shows highly significant results for all three pairings, although the low – high pairing has a W statistic of greater magnitude, suggesting even greater support for rejecting the null hypothesis. The effect size given is epsilon squared.

Thinking point

It is a commonplace truism that correlations (Chapter 6) do not prove cause and effect. Experimental and quasi–experimental results are still subject to interpretation.

Let us say that we replicated our study of alcohol and its effects on referees' judgement, with the same results. Do we know that this would hold for actual matches?

Validity, results meaning what we think they mean, is missing. To improve our insights, triangulation is wise, carrying out different types of investigation in order to view the phenomenon from a different perspective. Fresh insights can sometimes lead to a complete rethink.

In the study involving playing performance under investment regimes, we might look at conditions within the different regimes, perhaps to see if different stress factors are at work. We could examine such relationships as that between economic and political freedom, and whether or not other factors affect performance. See Chapter 6 for correlations and regression. Another type of research may involve interviews or focus groups; see Chapter 7 for how qualitative results are sometimes quantified.

This table of tests of difference is not exhaustive, but provides general guidance. **N.b.** *Non–parametric tests can be used with 'parametric' data, but in the author's view the reverse should not be happening.*

Design	Test	Conditions	Data
Same or paired subjects	Paired Samples T-Test	2	Parametric
	Wilcoxon signed rank	2	Non–parametric
	Repeated Measures ANOVA	3 or more	Parametric
	Friedman	3 or more	Non-parametric
Different Subjects	Independent Samples T-Test - Student's t	2	Parametric
	Welch's t	2	Parametric (good if equality of variances problem)
	Mann-Whitney U	2	Non–parametric
	ANOVA	3 or more	Parametric
	Kruskal–Wallis	3 or more	Non-parametric

Chapter 6 – Tests of relationships

Correlations

In this chapter, we are interested in the relationships between variables. Issues coming to mind include possible relationships between achievement and personal attributes such as motivation, between views on racial abuse and media coverage of the topic, and the possible link between the number of gambling websites and the incidence of match-fixing.

Statistically, the relationship is called **correlation**. The most used statistic is the **correlation coefficient**. It can run from 0 to 1, 0 being perfectly random and 1 representing a perfect positive relationship. It can also run from 0 to -1, -1 being a perfect negative relationship.

An example of a negative correlation could be the relationship between aerobic capacity and the time taken to finish a sprint.

As it is an important concept, let us dwell briefly on the non-causative nature of correlations. One variable may influence another, or it may not. The direction of causation may work in the opposite direction to what is expected. Or the relationship could be a horrible coincidence, with no causality whatsoever. But quite often, the explanation is elsewhere: there could be a mediating variable, one which explains the others. A famous example is the belief that night-lights may cause short-sightedness in children. This may be true, but results have been inconsistent. Another possibility is that short-sighted parents install night-lights; note of course that there is also a likely relationship between myopic children and myopic parents!

The following examples will give us a practical start, using the file Correlations.csv:

PerfPosA	PerfPosB	PerfNegA	PerfNegB
1	1	1	5
2	2	2	4
3	3	3	3
4	4	4	2
5	5	5	1

The first two variables in the Correlations.csv file have a perfect positive relationship with each other. The second pair of variables forms a perfect negative relationship. These examples do not occur in real life, but are here for demonstration purposes.

Press the Regression tab, select Correlation Matrix from the drop-down menu and transfer the relevant variables to the right-hand box. Keep the default options, including Pearson (to be discussed), but also choose Correlation matrix from the Plot section in order to see the scatter plot.

In this case, the two sub-samples have identical data: 1, 2, 3, 4, 5.

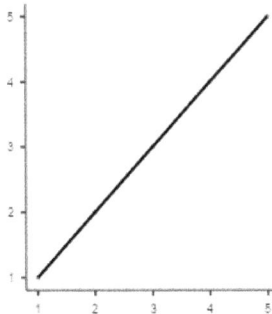

As well as producing a Pearson coefficient (r, or *rho*) of 1, the perfect relationship, with a very small p value, you can see this really neat scatter plot. A perfect correlation, $r = 1$, does not generally occur in research, but a slope moving from bottom left to top right generally indicates a positive relationship between two variables.

This is fine for showing what is going on, but for presentation purpose, I would suggest using Microsoft Excel's scatter plot, its equivalent in OpenOffice or LibreOffice Calc, or visualisation software.

Now place the two perfect negative variables in the right-hand box. Pearson's r should be - 1, again with a tiny p value.

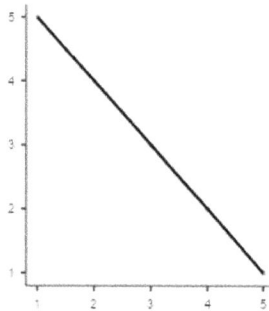

A slope rising from bottom right to top left indicates a negative relationship between two variables.

Now use the two randomly generated variables.

RandomA	RandomB
80	83
10	70
84	79
42	98
13	62
76	12
28	29
97	87
12	62
98	44

Note that the points representing pairings of data are scattered about the chart in a globular cluster. This is characteristic of the absence of a correlation.

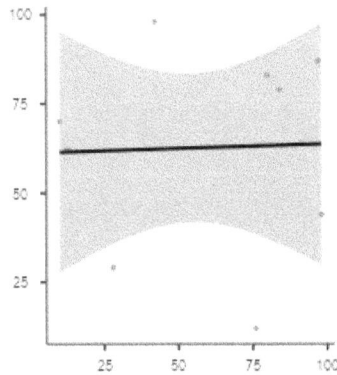

Pearson's *r* is 0.035, very close to zero. The *p* value 0.923 is very high, close to 1. In general, a low *r* denotes the absence of a relationship. Similarly, a high *p* value indicates lack of support for rejection of the null hypothesis; the null hypothesis is that there is no relationship between the variables.

Correlations and effect sizes

Let us consider a few key terms. The **correlation coefficient**, *r* (short for rho) in the case of the Pearson and Spearman tests, *tau* for Kendall's rank correlation, is a measure of the strength of the relationship in a particular direction (positive or negative).

$1 \Leftarrow 0 \Rightarrow -1$

A coefficient of 1 means a perfect relationship, -1 means a perfect inverse (or 'negative') relationship, with 0 as the ideal random relationship. So for a positive relationship between variables, the nearer to 1, the stronger it is; a coefficient of .924, for example, is clearly very strong in the positive direction. Similarly, a coefficient of -.924 would also be strong, but in the negative sense.

Before deciding that a positive correlation means a positive relationship between a pair or variables, or that a negative represents a conceptual inverse, it is well worth studying your data (as is recommended generally in this book). A common situation arises in questionnaire development, where scales run in different directions to avoid response order effects (for example, people always picking a high number regardless of the content of the question). The apparent direction of a relationship may be an artefact of the scales you are using.

The question of the size of a correlation is a vexatious one. (I cite positive sizes in the next two paragraphs, but you should consider negatives as the exact inverse.)

For smallish data sets, one could say that pairings of variables with a correlation coefficient of between .9 and 1 should be considered as very highly correlated; those of a magnitude of between .7 and .9 are highly correlated; those between .5 and .7 are moderately correlated; .3 to .5 are low correlations. Below .3, the relationship is weak or non-existent (Calkins 2005).

For larger data sets, it is more reasonable to consider anywhere between .5 and 1 as large, .3 to .5 as medium and .1 to .3 as small. It is also worthwhile looking at reports of similar studies and, when dealing with 'big data', effect sizes are likely to be more helpful than correlations or p values, as they deal with the variance.

The p value is our guide to how likely it is that the effect is 'real', that is, significant. The null hypothesis for correlations is that there is no reason to believe that a relationship exists; the lower the p value, the more likely it is that we can reject the null hypothesis. Sometimes, as in our perfectly negative and positive examples, you will see significance shown as something like $p < .001$. That doesn't mean that Jamovi has set a critical value (see Chapter 4); it means that the value of p is just too small to reproduce accurately.

The p value is particularly suspect when dealing with 'big data'. Lots of relationships will appear to be significant, often truthfully with that amount of evidence. The question then becomes just how meaningful are the relationships, and I would definitely extend this question to smaller data sets as well.

Coefficients and p values are sometimes mistaken for the strength of the effect. This is in fact represented by the **effect size**. A measure of the variance, this term is rarely seen in traditional books on statistical testing but is now increasingly viewed as important, particularly when dealing with big data. Essentially, one may be happy that a result is statistically significant, that an effect exists, but the effect size comments on the extent of its influence, its magnitude.

Sometimes you will come across Cohen's d, but there are a range of methods for calculating effect size. By default, for ease of use, I suggest the squaring of the correlation statistic. So if you have r from the Pearson test, you would calculate r squared (or r^2), multiplying r by itself. So if r is .73, the r squared is .73 × .73 = .53, just over half of the variance. Note that if we square a minus, in this case -.73, we still end up with .53; the strength of the effect is independent of direction. (Do test your calculator with the examples just given, as some calculators can't square negatives. If you choose to continue with a problem calculator and have to square negatives, you can just square positive values for the same result; .73 × .73 is the same as -.73 × -.73.)

There are no agreed reporting categories for effect sizes. Cohen (1977) recommends: Large: .8   Moderate: .5   Small: .2
If in doubt, examine the results of similar studies.

The Pearson test: a parametric correlational test

Participation	SelfConfidence
62	65
48	52
44	39
37	47
62	66
54	54
68	73
55	58
68	72
60	64

Groups of middle-school girls in 10 different areas are asked about their level of participation in sport and their feelings of self-confidence.

As previously, press the Regression tab, select Correlation Matrix from the drop-down menu. Transfer the Participation and SelfConfidence variables to the right-hand box.

The Pearson test is a parametric test and we need to check that the data meets the assumptions for such a test. The simplest thing to do is to go to the Plot section and choose both the Correlation matrix and Densities for variables. The former will provide the correlation chart, allowing you to check for linearity (of which more later) as well as to spot any outliers; to see the effect of an outlier, change an item of data to a large number and see the difference. The latter shows if you have something akin to normal distribution (I have used small samples, which make for less than beautiful viewing).

Another thing you can do is to open up the Paired Samples T-Test dialog and use the normality test there, the Shapiro–Wilk test (which is not significant in this example). Kim (2013) reports that formal tests such as Shapiro-Wilk are usable for samples smaller than 300, but may become unreliable for larger data sets.

Another option is to use the Exploration tab, opting for descriptive statistics, examining the Skewness and Kurtosis distribution statistics; the nearer to zero for each, the better. Between +1 and - 1 are perfectly reasonable figures. Although there are no official limits, many statisticians prefer the figures to be within the bounds of +2 and -2. West *et al* (1996) suggest a maximum of 2 for Skewness and 7 for Kurtosis. Kim (2013) recommends a limit of 1.96 for Kurtosis and Skewness for samples of less than 50; over 3.29 for samples of from 50 to 300; and for more than 300, go back to the correlation matrix plot and follow the advice of West *et al*.

With unsuitable data, we would need to use the Spearman or Kendall's *tau b* tests of correlation, as demonstrated in the next section.

As we have normally distributed data, we can carry on and use the default Pearson test. Select the Correlation matrix from the Plots section. A clear positive slope is to be seen and most of the data gathers quite closely to the slope.

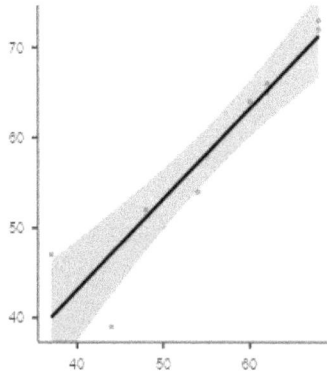

Pearson's r is .939, with an effect size of .882 (.939 squared), approximately 88% of the variance. This means that only about 12% of the variance from the mean is likely to be due to chance or additional factors.

If you select **Confidence intervals**, you get an estimate of the likely range of the correlation. We get the figures 0.757 and 0.986. This shows that about 95% of the pairings in this example are likely to have correlation coefficients between these figures. A wide spread suggests a lot of variability in the data, with a narrow spread suggesting greater confidence in the phenomenon (although note that large data sets can provide narrow differences between confidence limits because of minimization of discrepancies).

As with effect size, traditional textbooks have often ignored confidence intervals. You will find that some journals and indeed university departments state a preference for reporting effect sizes and confidence intervals as well as the traditional means and correlation coefficients.

Use the Exploration tab (for descriptive statistics) to check the means, 55.8 and 59.0, not very much different. Also, if you go to the Dispersion section of Descriptives and select Range, which tells you the difference between the highest and lowest figures in a data set, you see similar ranges. This suggests, but does not prove, a correlation. Again, these are useful statistics for reporting (as is Standard Deviation).

If you wish to examine the correlation against a one-tailed hypothesis, go to the Hypothesis section and choose Correlated positively in this case (in the aerobic capacity and race time example, you might choose Correlated negatively). If we choose the positive one-tailed hypothesis here, the confidence intervals will narrow to .803 and 1. Usually, unless your rationale is particularly clear, you choose the default Correlated (two-tailed) option.

An example where the null hypothesis cannot be rejected: variables NonSignA and NonSignB (return to the Correlated hypothesis setting).

NonSignA	NonSignB
52	60
53	34
47	38
40	52
48	54
45	55
52	36
47	48
51	44
38	56

Pearson's r is a minus number, -0.488, and the p value at 0.153 is quite high. The scatter plot shows much of the data deviating from the slope. This is a non-significant result.

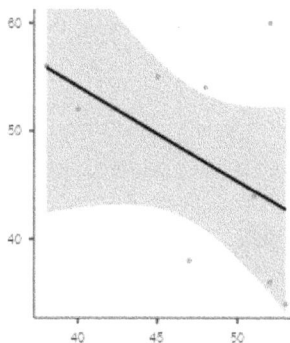

The Spearman and Kendall's *tau-b* tests: non-parametric correlational tests

Spearman is used in much of the traditional literature. These days Kendall's *tau-b* is becoming more popular. There is generally little difference in their interpretation of results, although it has been said that Kendall is more accurate for smaller samples (particularly less than 12) and ones with more tied ranks, while Spearman is better for nominal categories (for example, city 1, city 2 ... city x) with no hierarchical ordering. Do note that the statistical community (as usual) is divided over their respective qualities, but Kendall is now generally preferred. * Spearman's pervasiveness in much of the literature is because in the days before the personal computer, it was much quicker to calculate. It is also said by the cynical that the Spearman test is more popular because Spearman's *rho* is usually larger than Kendall's *tau*.

If your data set fails the Shapiro test, it is likely to have a non-normal distribution and is thus unsuitable for the Pearson test. In the Correlation Coefficients section, choose Spearman or Kendall's *tau-b* instead of Pearson.

Across ten different countries, averaged ratings are available of the confidence of disabled athletes in their countries' national sports bodies, confidence in the future of the sport and the level of financial investment in their sport (in 100,000s of dollars).

The scale for confidence in the sporting body (running at low ebb, it would seem) is:

1, 5, 4, 2, 2, 3, 1, 1, 3, 2

The scale for confidence in the future of the sport (uncertainty rules) is:

5, 4, 5, 3, 1, 2, 4, 3, 2, 2

The variables in Correlations.csv have been named GovtConf and FutureConf.

---

*You can't square Kendall's *tau* for effect size however.

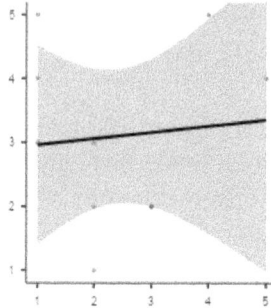

The Spearman coefficient is about $-.06$, obviously small, with a p value of 0.875, very big; any effect is clearly a matter of chance. Kendall's *tau b* readings are quite similar.

Now compare the confidence in the future (FutureConf) with this variable, Investment (representing thousands for each area):

28, 30, 25, 27, 18, 20, 15, 24, 18, 22.

Chapter 6 – Tests of relationships

Let us assume that we were inclined to expect investment to be positively associated with confidence in the future of the sport. We therefore select the one-tailed hypothesis, selecting Correlated positively from the Hypothesis section and, for clarity, Flag significant correlations.

Correlation Matrix

		FutureConf	Investment
FutureConf	Spearman's rho	—	
	p-value	—	
	Kendall's Tau B	—	
	p-value	—	
Investment	Spearman's rho	0.572 *	—
	p-value	0.042	—
	Kendall's Tau B	0.483 *	—
	p-value	0.032	—

Note. H$_a$ is positive correlation
Note. * p < .05, ** p < .01, *** p < .001, one-tailed

As we consider the less rigorous one-tailed level to be acceptable on theoretical grounds, and have chosen the critical value of $p < .05$, then we can reject the null hypothesis.

If we had chosen the two-tailed hypothesis, using merely Correlated in the Hypothesis section, then we would have seen the p values doubling, losing our asterisks.

If you select Correlation matrix from the Plot section, you will see a mild slope.

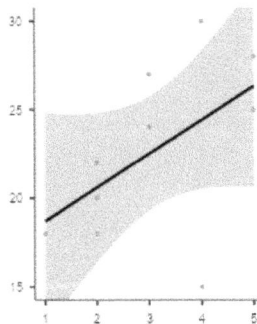

A cautionary note: It is a very good practice to examine a scatter plot as well as using a test. The reason is that tests of correlational relationships, non–parametric as well as parametric, are linear. A relationship essentially runs in a straight line. Assuming significance or non-significance without using a scatter plot could be a big mistake. Let us look at two examples.

One error was made by the author (reader gasps). I examined a friend's blood pressure readings against the time of day (thanks, you know who, for permission to reproduce this evidence of my impulsive nature). We had expected a relationship between the two variables and were surprised to find a non-significant result, with a coefficient of –0.16 and with .3 for a two–tailed p value. Then, it dawned on me.

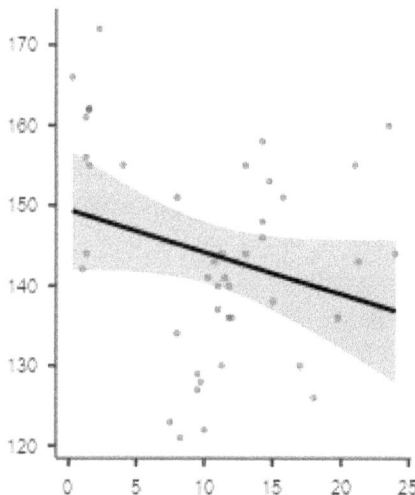

This expanded image of the chart shows something like a W shape in the data far away from the slope. My friend tended to have higher blood pressure readings in the early afternoon and at night. There is a pattern, but the effect is not linear. A test for correlation was of course meaningless, as was any projected slope.

Now, for students under stress as well as sports fans, I present a classic non-linear relationship, between performance and arousal:

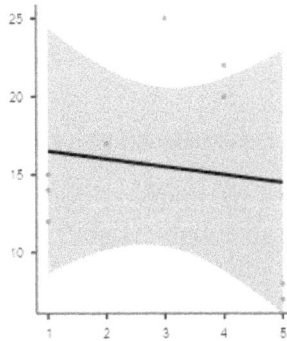

The loose data appears as an inverted U shape. This is a crude example of the Yerkes-Dodson Law (Yerkes and Dodson, 1908). Some degree of stimulation appears necessary for performance, but high stress and low levels of arousal may damage performance. The non–linear effect emerges from the chart, but correlational tests, parametric or non-parametric, would be inappropriate. These tests are not appropriate for 'curvilinear' relationships. You should always examine your data visually before considering test results.

If we return to our main parametric example with a significant result, the relationship between sports participation and girls' self-confidence, a correlation coefficient of .939 when squared gave .882, a large effect size. Now consider the relationship between the variables indicating Satisfaction with life and Income (fictional data).

Correlation Matrix

		Income	Satisfaction
Income	Pearson's r	—	0.725**
	p-value	—	0.009
Satisfaction	Pearson's r	0.725**	—
	p-value	0.009	—

Note. H_a is positive correlation
Note. * p < .05, ** p < .01, *** p < .001, one-tailed

We have reason to reject the null hypothesis, that there is no relationship, with a p value that is smaller than .01, but squaring the coefficient gives us an effect size of .53, not much more than half of the variance. One way of examining a phenomenon more broadly is to look at correlations of more than a single pair of variables.

Multiple correlations – parametric - using Pearson's test

Thus far, we have focused on relationships between two variables. It is quite possible to use several variables, as we do here. Do note, however, that the more correlations you include, the more likely that some apparently significant relationships are in fact due to chance (Type 1 errors). So try to have a rationale for including each variable.

On this occasion, we are interested in a (fictional) study of obesity among teenagers. Our particular interests are the connection between excessive television-watching and obesity, and the possibility that recreational drug-taking is positively related to obesity. As well as measures of drug-taking, television watching and obesity, we also include measures of educational attainment and economic status.

Press the Regression tab, select Correlation Matrix from the drop-down menu and transfer the relevant variables to the right-hand box.

If you choose the Correlation matrix from the Plots section, you will see an excellent grid of scatter plots for each pairing. For the correlation matrix itself, let us assume that you have started with the default two-tailed hypothesis, Correlated, without a predicted direction. The Flag significant correlations option makes it easier to detect significant results; asterisks appear on the table of results.

Correlation Matrix

		Education	Economic	Drugs	Obesity	Television
Education	Pearson's r	—				
	p-value	—				
Economic	Pearson's r	0.743 **	—			
	p-value	0.009	—			
Drugs	Pearson's r	-0.218	-0.260	—		
	p-value	0.520	0.440	—		
Obesity	Pearson's r	-0.405	-0.634 *	0.492	—	
	p-value	0.217	0.036	0.124	—	
Television	Pearson's r	0.355	-0.025	0.332	0.600	—
	p-value	0.284	0.941	0.318	0.051	—

Note. * p < .05, ** p < .01, *** p < .001

It is clear that educational attainment and economic status are highly correlated, with a moderate negative correlation for economic status and obesity. In both cases, they are considered significant. This is not the case when we consider the relationships of interest, although the pairing of television watching and obesity is moderately correlated, with a lowish correlation pertaining to drug-taking and obesity, and also a low negative correlation between education and obesity.

Now we need to make a decision about the correct use of one-tailed hypotheses. This needs to be made before taking a peek at other options in the Hypothesis section. Our concern combines the theory (or rationale) with statistics: Would one have expected a particular direction for the pairing? If yes, then a one-tailed test is reasonable; if no, then it is not. The one reason which is not acceptable is that you want a significant result.

Correlation Matrix

		Education	Economic	Drugs	Obesity	Television
Education	Pearson's r	—				
	p-value	—				
Economic	Pearson's r	0.743 [**]	—			
	p-value	0.004	—			
Drugs	Pearson's r	−0.218	−0.260	—		
	p-value	0.740	0.780	—		
Obesity	Pearson's r	−0.405	−0.634	0.492	—	
	p-value	0.891	0.982	0.062	—	
Television	Pearson's r	0.355	−0.025	0.332	0.600 [*]	—
	p-value	0.142	0.530	0.159	0.026	—

Note. H_a is positive correlation

Note. [*] p < .05, [**] p < .01, [***] p < .001, one-tailed

Our rationale for the study in this particular milieu was that obesity and television watching are likely to be positively related, so at least for this relationship, we consult the table using a unidirectional hypothesis, Correlated positively.

Do note that when you look at the tables through the lens of a particular hypothesis, it will apply to all of the pairings, which obviously does not make sense. So when reporting, it is best to report each pairing separately, referring to 'one-tailed' or 'two-tailed' in each case.

As we can see, Obesity and Television watching are considered to be significantly related. Although whether or not the null hypothesis should have been rejected in the previous matrix because of a p value of 0.051 is of course questionable. [*]

---

[*]The critical value of $p < .05$ was of course Fisher's rule of thumb. You may want to study this data using Bayesian analysis, which gives a graded set of reported results rather than a 'significant or non-significant' result; see Chapter 16 and Chapter 17.

Multiple correlations – non-parametric
- using Spearman/Kendall's *tau-b*

Press the Regression tab and select Correlation Matrix from the drop-down menu. Transfer the relevant variables to the right-hand box.

The interest here is in male professional soccer players' attitudes to foul play, using uncalibrated 5-point Likert scales. If you checked normality by going to Paired T-Tests with these variables, you would find that most do not fail the Shapiro-Wilk test (that is, the results are not significant). However, I am wary of treating scales as if they comprised continuous data. The basic rule is that if the concept can't really be halved, for example that half of a 4 rating really is not equivalent in meaning to a 2, then it should not be treated as truly continuous.

There are other reasons why you might not want a parametric test. Perhaps the variables are of differing types of data. Perhaps you do not think that they are close to normal distribution because of the graph or skewness or kurtosis. Sometimes, as I said earlier, I think you just have to look at the type of data you are putting in and make a logical decision.

I have opted for the Flag significant correlations option, which will show us asterisks.

Correlation Matrix

		Authoritarianism	Empathy	Integrity	Fouls
Authoritarianism	Spearman's rho	—			
	p-value	—			
	Kendall's Tau B	—			
	p-value	—			
Empathy	Spearman's rho	−0.233	—		
	p-value	0.491	—		
	Kendall's Tau B	−0.172	—		
	p-value	0.507	—		
Integrity	Spearman's rho	−0.341	0.208	—	
	p-value	0.304	0.540	—	
	Kendall's Tau B	−0.211	0.170	—	
	p-value	0.413	0.509	—	
Fouls	Spearman's rho	−0.228	−0.776**	−0.190	—
	p-value	0.500	0.005	0.575	—
	Kendall's Tau B	−0.200	−0.697**	−0.176	—
	p-value	0.445	0.008	0.499	—

Note. * p < .05, ** p < .01, *** p < .001

The correlations table only picks up on one relationship, a negative one between fouls and empathy. (This is very loosely based on Sezen-Balcikanli and Sezen 2017.)

We can feel reasonably sure that we can reject the null hypothesis that the effect does not exist, but how important is the effect in terms of real life? If we square Spearman's *rho* (-.776 × -.776), we get an effect size of .602, 60% of the variance.

Various ideas emerge from this. If the rest of the variance is random 'noise', how can our research model be improved? Is there another influential variable which would help us to better understand the effect and create a better model? Is the new model relevant and should we invest time and resources into it? These questions become clearer when we consider multiple regression.

As a building block, however, we first need to consider regression as a concept.

Regression

In general terms, regression is a predictive tool used to create statistical models. A model is not intended to be an accurate reflection of the world, but an idealization. It creates a usable set of concepts to help us handle a problem or theory in a practical way. Multiple regression is a major tool in this process, but in order to understand multiple regression, we need to consider the concepts within simple linear regression. We will also be building upon our knowledge of correlations.

Simple linear regression (two conditions) – parametric

Regression appears on the face of it to be similar to correlations. We are still interested in the nature of relationships between data and we are still interested in a slope, although this is known in regression as *the line of best fit*. Both methods assume a normal distribution, although arguably this is more important for the dependent variable in regression. Both require linear relationships.

There are some differences. The mathematical method is different. Although I will not cite the equations, you do need to understand what regression does, and with what type of variable. We will examine it here by comparing it with the already familiar concept of correlations.

Correlations examine a mutual relationship between variables, with no mathematical differentiation between the variables. We only know that they are related to each other. Any choice of direction, with a one-tailed hypothesis, is a matter for the test user; any assumption of cause and effect is at the researcher's own risk.

Linear regression assumes that one variable affects another, assuming one direction (if you ever go to another book, you will find two formulae, one for how x influences y and another for y influencing x, but you don't have to go there). So the dependent variable, the one being affected, needs to be a continuous variable which can be acted upon. For example, we might want to consider salaries as a dependent variable, or the number of incidents of a particular nature, or match scores. One relevant question of this type is whether or not the number of training sessions in a season affects performance. Another is whether or not levels of policing at stadia have an effect on hooliganism. Yet another would be the effect of wall-poster promotions on attendance at matches.

One important feature of regression is that the relationship between the two variables can lead to predictions. When we have a line of best fit in simple regression, which involves only two variables, this can be extended to extrapolate beyond the line. Numerically, we can see how far one unit of the predictor (referred to as the covariate in Jamovi) produces an increase in the criterion variable (dependent variable). For example, the regression could indicate that for every additional training session, goals rise by a certain amount (or fall?). For each extra police officer, so many fewer fights occur (fictional data as usual). In other software and some of the literature, the predictor is sometimes known as x (as in the horizontal axis of a chart) and the dependent variable as y (the vertical axis).

As usual, some common sense is required. In the case of policing, it is quite likely that the number of incidents affects policing levels as well as vice-versa. So although regression is used with the aim of describing cause and effect, it still needs to be applied logically.

Let us look at a practical example. Use the Regression.csv file.

Firstly, it helps to check that there are relationships between the data. If you open up Regression / Correlation Matrix and examine the variables Education and Income with the correlation plot, you will see a clear, strong, linear relationship. If our data refers to the educational level of individuals and their subsequent earnings, then it is clearly worth investigating Education as an explanatory factor, influencing dependent variable Income.

However, if Income means parental income, more likely to be a causal variable, we may need to reverse our variables, making Education the dependent variable – if you experiment with the correlation matrix plots and reverse the order of the variables, you will find a subtle difference in the scatter plots. Whereas the correlation coefficients are not affected by the order in which the variables are placed, the plot is different, which is important when applied to regression. Regression assumes causation, unlike correlations.

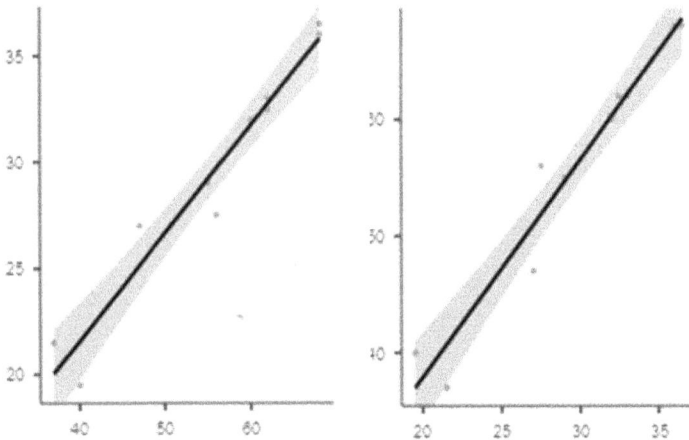

Press the Regression tab, selecting Linear Regression from the drop-down menu. Income fits into the Dependent variable box, with Education transferred to Covariates.

Linear Regression

	Dependent Variable
Status	Income
SalesA	
PriceA	Covariates
Sales	Education
Price	
Instore	
Street	
Radio	
	Factors

	Education	Income
1	62	32.5
2	56	27.5
3	40	19.5
4	37	21.5
5	62	33.0
6	47	27.0
7	68	36.5
8	55	29.0
9	68	36.0
10	60	32.0
11		

Do note that Income is in thousands (e.g. 32.5 = 32,500). Education is based on a rating.

Linear Regression

Model Fit Measures

Model	R	R²
1	0.973	0.947

Model Coefficients

Predictor	Estimate	SE	t	p
Intercept	1.208	2.4009	0.503	0.628
Education	0.509	0.0425	11.966	< .001

The default 'Model Fit Measures' for regression models in Jamovi are the coefficient R and what we would normally call the effect size, R squared. In regression, R squared is often known as the coefficient of determination, the proportion of variance in the dependent variable which is predictable by the independent variable(s). As usual, this ranges from 0 to 1. An effect of the magnitude of our fictional example is rare and unlikely in this particular scenario.

The idea of model fit is how well the observed data fits with the model, in this case the idea that education affects income. The Model Fit Measures table will assume greater importance when we test out different models using multiple regression (which is also when we use Adjusted R squared and other measures).

In the Model Coefficients table, the first thing to look at is the p value. As usual, we want a low p value to tell us that we have significance or, correctly, that we have evidence to support the rejection of the null hypothesis of no causal effect.

We then see how much of the dependent variable is predicted by each unit of the explanatory variable. The relevant statistic is the 'Estimate' (coefficient estimate). Here, you have to get your head around the idea that the figure referring to your predictor (or independent variable), here Education, actually refers to its influence on the dependent variable (Income). If we were dealing with single units, we would simply say that for every *one unit* increase in the education rating, income would increase by .509 units (half a unit). However, as our expenditure is really in thousands, we say that Income = 0.509 × 1000, giving $509. So every increase in the education rating is, on average, predictive of $509 in income.

Before leaving simple linear regression, let us consider the sales of a monthly gym membership, working out the effects of prices on the number of sales at various sports centers. This example will be expanded when we cover multiple regression. The variable PriceA is the predictor (covariate); SalesA comprises the dependent variable. In general, the lower the prices, the higher the number of sales.

Model Fit Measures

Model	R	R²
1	0.553	0.306

Model Coefficients

Predictor	Estimate	SE	t	p
Intercept	15068	2268.1	6.64	< .001
PriceA	−261	90.2	−2.90	0.009

The proportion of variance is about 30%, with p = 0.009. As the result is significant, we can consider the Estimate.

Note that the coefficient estimate for PriceA is negative, reflecting the inverse relationship between price and sales. It is always worth checking that the variables make sense; if they don't, it may be that a key variable has been left out of the model.

Before we consider other factors, it is worthwhile checking that the data are reliable. (Some of the data used here is not from a normal distribution as the author wished to achieve various effects with a small data set.) Let's visit the Correlation Matrix and check out the plot for SalesA and PriceA.

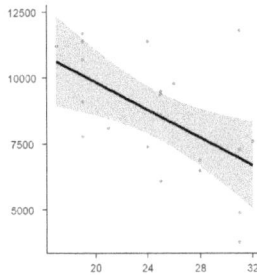

Note that we generally put the predictor (price) on the horizontal (x) axis and the dependent variable (sales) on the vertical (y) axis. Well, the relationship is linear, the bottom right to top left indicating a negative correlation, but there is a problem: an outlier on the top right of the chart (coordinates approximately 30, 12000). This is why it is advisable to look at the data with a graph before focusing on tests. One coordinate is quite remote and is theoretically dubious: one center is selling at the highest price and yet is also selling well. It could be a really exclusive area, but let's not go there.

Usually in statistical testing, removing real information for convenience is unforgivable (unless they are input errors). However, we are building a predictive model. In such a case, it is acceptable to remove the outlier to improve the model. We want to study usual behavior.

Use the emended variables Sales and Price. These have omitted the outlying pairing (the last pair in the SalesA and PriceA variables, should you wish to examine the data).

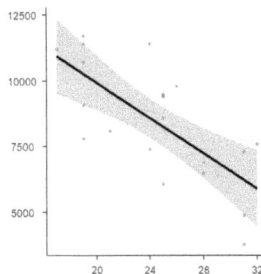

Note the more tightly knit pattern. Let us now update Linear Regression with the Dependent variable Sales and the predictor (covariate) Price.

Model Fit Measures

Model	R	R^2
1	0.715	0.511

Model Coefficients

Predictor	Estimate	SE	t	p
Intercept	16628	1916.1	8.68	< .001
Price	−335	77.2	−4.34	< .001

We now have R squared = 0.511, a big improvement. However, 0.511 still accounts for only half of the variance. Before we worry about that, let's do our little explanatory sum. As we are not dealing with thousands, but just single units, we can simply write that for every increase of a dollar in price, sales fall by, on average, about 335. Assuming that you want to escape from using the y and x beloved of many statistics books, you would write this formally as, Sales = +16628 -335 (Price). The first figure is the coefficient estimate for the Intercept, where the line of best fit passes through the y (dependent variable) axis.

Another statistic is the F test, to be found in the Model Fit section. The larger the F ratio statistic, the greater the variation in group means. In this example, F is 18.8, a fairly substantial figure.

It is likely, however, given the proportion of the variance, that additional variables may provide a more explanatory model. We will try this with multiple regression.

Standard (simultaneous) multiple regression – multiple predictors against one dependent variable (parametric)

Returning to our example of sales of monthly gym memberships, we may ask if price is the only significant factor in determining the number of sales. Multiple regression allows us to build a model for effective prediction. We are particularly interested in two issues:
Will additional variables make an appreciable difference to predictions? If they do, are some variables more useful than others?

Press the Regression tab and select Linear Regression from the drop-down menu. Transfer Sales to the Dependent Variable slot, with the following variables to the Covariates box: Price, Instore, Street and Radio. As well as being interested in how far price affects sales, we are also interested in the contributory effects of promotion methods: are sales affected by promotions within the centers, street advertisements and on the radio? (Remember to use Sales and Price, not our rather less tightly knit variables SalesA and PriceA.)

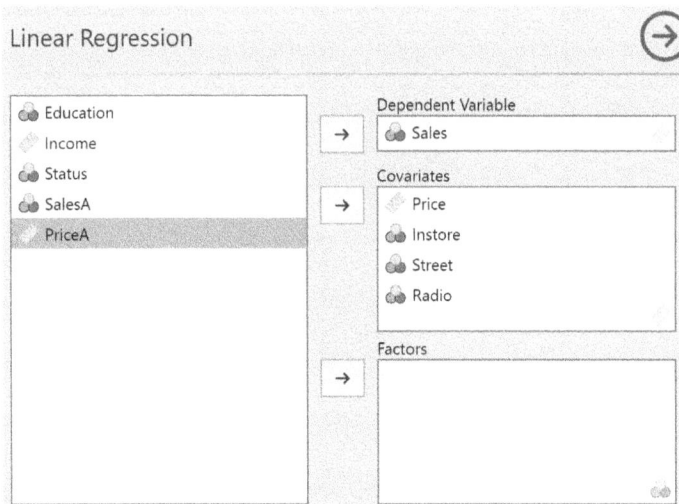

There is also a Factors box. This means that you can also enter groupings, for example gender or class. You can have both covariates and factors at the same time, but do note that as usual, the more complicated the input, the more difficult it is to interpret the results.

Model Fit Measures

Model	R	R^2	Adjusted R^2	F
1	0.878	0.770	0.709	12.6

As our model reflects the complexity of the real world, the regression coefficient R is quite a lot bigger than the simple linear regression model of Price and Sales, with a much stronger R squared. However, Adjusted R squared, as found in the Model Fit section, is preferred with multiple regression; this takes into account the number of predictors and gives a more conservative effect size than R squared. Our effect size is a quite impressive 0.709, about 70% of the variance.

The F test result is 12.6. This is substantively smaller than the statistic when Price was a lone predictor and suggests that not all of the variables are contributing much to the variance. We can be more specific when we look at the coefficients for the different variables.

Model Coefficients - Sales

Predictor	Estimate	SE	t	p
Intercept	-2105.259	4788.596	-0.440	0.666
Price	-301.894	58.730	-5.140	< .001
Instore	1.384	1.235	1.121	0.280
Street	0.472	0.348	1.356	0.195
Radio	0.944	0.289	3.265	0.005

In addition to the inverse presence of Price, only Radio promotion appears to contribute significantly to the variance.

So our first question is answered positively in this instance: additional variables have made a considerable difference to the predictive model. It also seems likely that one of our additional variables, Radio, is more important than the others.

Before finding out about the extent of influence of the two main explanatory variables, it is sensible to check that we have met the statistical assumptions for multiple regression. Our predictions rely on the fact that the assumptions for the model hold.

As mentioned before, we need to check that we have normally distributed data and that the data is linear. We also need to check that the errors between observed and predicted values – the residuals – are normally distributed. Yet another problem is too much multicollinearity, over-strong relationships between the variables. A further problem is autocorrelation, over-strong relationships between the residuals.

Let us first consider residuals. Within the Assumptions Checks section, choose Q-Q plot of residuals and Residual plots.

Q-Q Plot

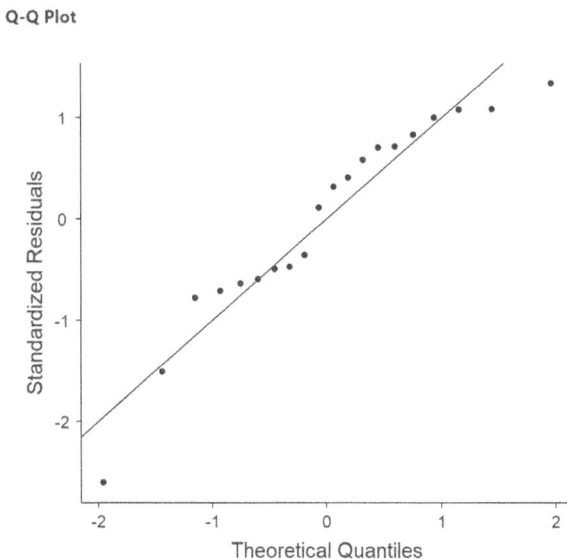

The residuals cling quite closely to the straight line, so we're happy with the Q-Q plot.

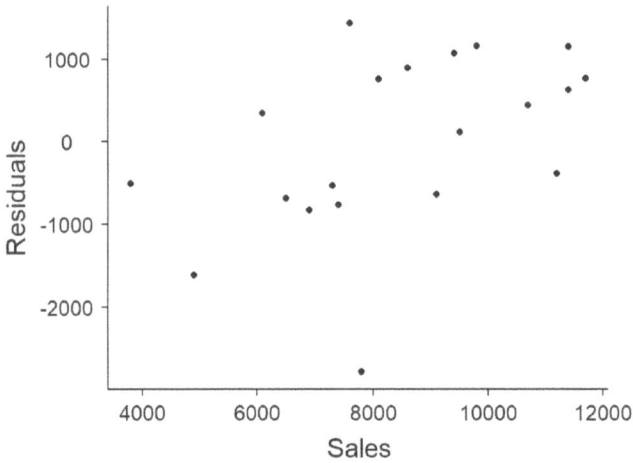

We are particularly interested in the dependent variable, Sales in this case. The residuals are quite randomly spread, which is fine. If the data formed a concentrated ball around the horizontal line, this also would not be a problem. What you do *not* want is either a U shape or its inverse, a rounded archway shape; these indicate curvilinear relationships.

Next we look at the collinearity and autocorrelation options.

Collinearity Statistics

	VIF	Tolerance
Price	1.03	0.975
Instore	1.21	0.828
Street	1.18	0.849
Radio	1.08	0.929

High collinearity indicates over–correlated variables, which may mean that they are measuring the same underlying construct. As suggested elsewhere, throwing in a load of variables and hoping for something to come up is not a good idea; just include variables that make sense. We want a low VIF – variance inflation factor – not much above 1; these figures, between 1.03 and 1.21, are fine. We want a high Tolerance factor, approaching 1; the Tolerance column here is good, between .828 and .975.

Durbin–Watson Test for Autocorrelation		
Autocorrelation	DW Statistic	p
−0.276	2.49	0.244

The Durbin-Watson statistic approaching 0 is a positive autocorrelation. If it approaches 4, there is a negative autocorrelation. The score should be between 1.5 and 2.5, as our result is, although some leniency could be shown as far as 1 and 3 respectively. However, it seems reasonable to use the p value. The null hypothesis is that there is no autocorrelation between the residuals. The result above does not support rejection of the null hypothesis; we're ok. (But do note that the p value will be slightly different each time the test is run.)

Another point to consider is sample size; results tend not to generalize with small samples. Stevens (1996) recommends 15 participants per predictor. Tabachnick and Fidell (2007) recommend the following formula: 50 participants + (8 x the number of predictors/covariates). So with 4 predictors, the minimum acceptable sample according to the former would be 4 x 15 = 60 participants, with the latter recommending 50 + (8 x 4) = 82. Given the small size of my fictional sample, it is unlikely that all of this would be replicated.

Returning to our results, we can run the multiple regression with only the predictors Price and Radio. This accounts for considerably more of the variance than our simple linear regression model using Price. The figures for the simple model were $R = 0.715$, R squared = 0.511, Adjusted R squared 0.484, $F = 18.8$. Here is the model summary using the Price and Radio predictors.

Model Fit Measures

Model	R	R²	Adjusted R²	F
1	0.833	0.694	0.658	19.2

The model summary for Price and Radio gives us considerably larger coefficients R and Adjusted R squared, the latter indicating two-thirds of the variance, compared to about half covered by the simple model. At the same time, there is a limited decline from the figures for the more comprehensive set of predictors (Adjusted R squared = 0.709). So, assuming that I have not omitted another useful predictor, we probably have the optimal model, explaining the sales figures via only two predictors while not drastically reducing the variance. The statistical term for explaining or predicting with as few explanatory variables as possible is a **parsimonious model**.

You may also want to use the AIC, BIC and RMSE statistics to compare models. These measures of model fit are illustrated in the next section, on hierarchical regression, but they can also be used within this approach.

Model Coefficients - Sales

Predictor	Estimate	SE	t	p
Intercept	3669.653	4362.780	0.841	0.412
Price	−303.077	63.668	−4.760	< .001
Radio	0.973	0.306	3.181	0.005

Now we are in a position to estimate the effects of price and radio promotion on sales. As usual, ensure that the unstandardized coefficients make sense in relation to the model. As we are not dealing with thousands, as in our income example, but just raw numbers, we don't have to multiply anything. Sales decrease by 303, on average, for every dollar added to the price (note the negative coefficient) and there is almost one sale (.97) for each dollar spent on radio promotion. Formally, this can be written as Sales = 3669.653 -303.077(Price) +0.973(Radio).

The procedure we have followed above is called standard multiple regression, as it is the most frequently used, or simultaneous multiple regression, all variables being examined at the same time. Other approaches to multiple regression order the entry of the variables.

Hierarchical regression

The Model Builder section allows us to examine the predictors in blocks (also referred to in the literature as 'steps'), specifying the order in which the predictors are introduced into the multiple regression equation. On theoretical grounds, or at least with a clear rationale, we can compare successive regression models, arguably allowing a more principled way of examining the inter-relationships. Variables can be processed in individual steps, although it is possible to put a set of closely related variables into the same block. What we want to know is whether or not there are changes to predictability over and above the preceding variables.

A common use of hierarchical regression based on research relevance is when there are variables already known to be predictors. These would be put in the first block. 'New' variables would then be placed in the second block to see if they improve the model.

Another typical, and quite similar, usage is to put demographics into the first block. The second block would contain variables already recognized to be of significance, thus replicating previous research, with a third block containing the variable of interest in the current study. Let us say, for example, that we are interested in problem gambling and peer and parental conflict. The first block would perhaps contain age and gender. The second block could comprise measures of impulsiveness and attitudes towards gambling. The third block could derive from sources of conflict surrounding the individual.

Yet a further use is temporal. For example, gender would (usually) precede a set of attitudes.

Returning to hierarchical regression in general, each new predictor ('covariate') is assessed for its effect on the criterion (dependent variable), while the equation controls for the previous predictors. The relative contribution of each block is assessed, as is the overall model.

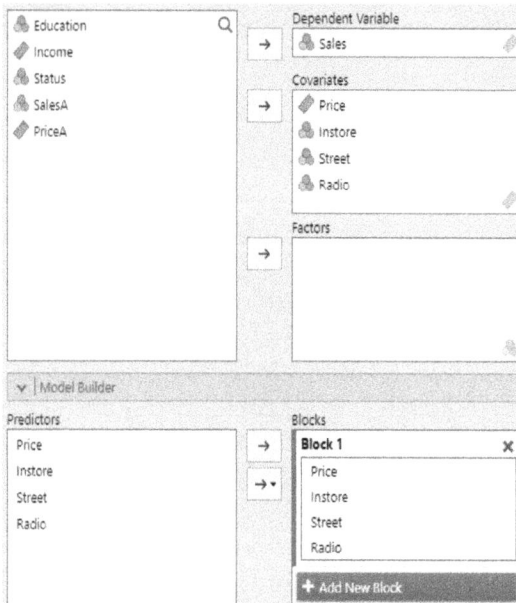

To start with, put all the relevant variables into the relevant boxes at the top and open the Model Builder section.

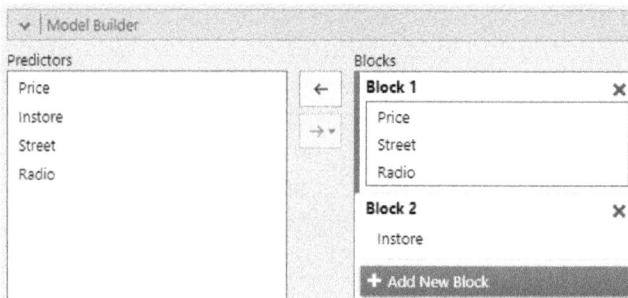

Chapter 6 – Tests of relationships

Price, Street and Radio promotion are all well-established predictors of Sales. We are interested in whether or not in-store promotions are effective in creating sales. Press the 'Add New Block' bar and drag the predictor of interest into the new block.

Go to Model Fit and select all of the measures except for R and Adjusted R squared. R squared is a very important measure in hierarchical regression. AIC, BIC and RMSE are popular for model comparisons.

Model Fit Measures

Model	R^2	AIC	BIC	RMSE	F
1	0.751	346	351	1085	16.1
2	0.770	347	353	1042	12.6

Model 1 represents Block 1, here containing the established predictors. Model 2, adding Instore, includes the previous block, but controls for the previous variables.

R squared is an effect size that can take values from 0 to 1. A higher figure means that more of the variance is explained by the model. This is always reported in hierarchical regression. Here, this is not much larger for Model 2 than for Model 1, generally indicating that Model 2 has poor explanatory power. Other measures tell a similar story. Let's look at the F statistic on the right: there has been a fairly sharp drop in the size of the F ratio when we reach Model 2. (Ignore the significance test for 'Overall Model Test', to the right of the F ratio but not shown here; it invariably gives a significant result.) Such a drop indicates a less parsimonious model. The F statistic is also reported.

The Akaike information criterion (AIC) and Bayes information criterion (BIC) are both criteria for testing the quality of relative models, looking for parsimony. A parsimonious model is one which has useful explanatory value with a limited number of predictor variables. Unlike R squared and F, drops towards smaller values are good signs for the information criteria. In our example, there is a small rise in values from Model 1 to Model 2, indicating that the most comprehensive model has less explanatory power.

AIC and BIC often come out with similar results but as they work on different assumptions, this is not always the case. On the occasions where they vary, cite both statistics in your reporting and consider the range of models indicated as potentially useful by both. As usual in applied statistics, you should consider each model rationally on its merits.

RMSE (root mean square error!) is an absolute measure of fit, which has as its advantage that it uses the same units as the dependent variable. Models with a smaller RMSE are indicated as having a better fit, but while RMSE tells you more about how well models fit the data, it is less easily interpretable than R squared. In our case, the RMSE is smaller for Model 2 than Model 1, but we need to consider this proportionally: a drop of 43 when we are talking about figures of over 1000 is not particularly impressive.

Model Comparisons

Comparison						
Model	Model	ΔR^2	F	df1	df2	p
1	- 2	0.0193	1.26	1	15	0.280

Delta R squared (delta being represented by a triangle) is the difference between the R squared statistics of two models. This should be reported, alongside the R squared and F from the Model Fit Measures table. To the right of Delta R squared, we see the results of the Model Comparisons test: there is no clear difference between the first and second model, so our variable of interest, as represented in Model 2, does not seem to make a difference.

Model Specific Results [Model 1 ▾]

Model Coefficients - Sales

Predictor	Estimate	SE	t	p
Intercept	−878.866	4699.016	−0.187	0.854
Price	−303.240	59.185	−5.124	< .001
Street	0.621	0.324	1.916	0.073
Radio	1.011	0.285	3.547	0.003

This table shows the nature of hierarchical regression, one model within another. The table shown first (but not shown here), Model 2, contains all the variables and has the same results as our simultaneous multiple regression. One interesting point shown in Model 1 is that the Street variable has a rather smaller p value than in the more comprehensive model. While with a small sample like this, a fluke is quite possible, it can also be that the better fit of Model 1 might have this effect. Do note from a worldly perspective that the possibility of 'significance' does always mean a meaningful effect; one dollar spent on street advertising appears to produce an increase of .621 of a sale on average and even radio only approximates a single sale per promotion unit.

Which type of multiple regression should I use?

When there is no reason to put any particular variable in before another one, or you are conducting largely exploratory work, or the logic of what you are doing is unclear, then simultaneous regression is the most appropriate method. Many find it safer, hence 'standard regression'.

For more theoretically based work, hierarchical regression may be preferred, as it should add to your understanding. I must stress that a clear, explicit rationale is required for this approach. It is really for small numbers of variables. Pedazhur and Schmelkin (1991) recommend no more than three variables and even then consider this approach, which they refer to as variance partitioning, as "of little merit.. no orientation has led to greater confusion and misinterpretation of regression results.."

Thinking point

One problem with correlations, already alluded to, is that too many variables tend to result in fluke correlations. Another weakness of correlations is that they do not prove cause and effect. For example, even if we are sure of a reliable relationship between perceptions of police effectiveness and crime reduction at sports grounds, are we sure that perceived effectiveness reduces perception of crime risk? Or could it be that fear of crime influences individuals' attitudes to the performance of the police? Or should we take into account a mediating factor such as what respondents see on television?

Running experiments or quasi–experiments is one way of dealing with this problem. When this is not practicable, we could triangulate: different aspects of a problem may be subjected to different forms of analysis, perhaps using different methodologies, to find out if the original theory can be disproven. Confirmatory factor analysis could be used to test the current model or to seek other explanatory models. Methods such as structural equation modeling can also examine the direction of correlational effects.

This table of tests of relationships is not exhaustive, but refers to tests cited in this chapter.

Purpose	Data	Number of variables	Test
Correlation	Parametric	2 or more	Pearson
	Non-parametric	2 or more	Spearman or Kendall's tau-b
Prediction and modeling	Parametric	2 or more predictors	Linear regression (standard and hierarchical)

Chapter 7 – Categorical analyses

Introduction

Speaking categorically

In this chapter, we are not interested in measurable data. We are interested in counts of observations, otherwise known as frequencies. In each example, the data fits into **exclusive and exhaustive** categories: each observation may only be included in just one category and the total number of categories must contain all of the observations in the study.

Let us say that for the purposes of a survey, qualified sports teachers must fit into the categories of Employed, Unemployed and Retired. Each individual observation must go into just one of these categories; so exclusiveness means that a decision needs to be made about in which category to place a retired teacher who is occasionally hired, or a casually employed teacher. Exhaustiveness means that no sports teachers in the survey remain outside of the categories.

In practice, this means that you have to make some subjective decisions. In the above case, you might decide on a further category, maybe Mixed, if a substantial number of teachers fall into more than one category. In other cases, you may wish to subsume categories into broader ones.

Angry	Irritated	Neutral	Positive
6	33	38	23

If you have this range of attitudes derived from a set of 100 feedback forms about a plan for a new sporting venue, you may wish to let the research commissioner know about the extreme negativity of the six individuals and the overall reasons. However, this type of statistical test will be heavily influenced by any discrepancy such as a really small (or large) number.

Given that the very small Angry category will certainly lead to a significant, and obvious, result, it would make sense in terms of the statistics to concatenate the Angry and Irritated categories into one category (Negative, for example). The decision to combine two or more categories is a subjective one.

This and the absence of measurement means that such statistics are often called **nominal**. A strong rationale needs to be balanced with statistical considerations. Subsuming a group of categories, for example, needs to make sense.

The quantification of qualitative data

I often come across the assumption that qualitative research is something completely unrelated to quantification. Some of its proponents may even say outright that the results of their interviews or focus groups are not measurable. Without getting personal, I will suggest a few reasons why you would want to count, for example, the number of interviewees who think in a particular way about a subject or the number of focus groups changing their minds over a topic.

You might worry that the forceful viewpoint of a few respondents is clouding your judgement, and that many would not agree. You might also want to see if particular levels of status are more prone than others to share a particular view. Or you might want to look at the prevalence of a particular idea in comparison with conflicting ideas.

What's happening, statistically speaking?

At the core of these tests is a very simple principle: We are contrasting the **observed frequencies** with the **expected frequencies**.

The observed frequencies are the actual numbers that you have within each cell of the table, for example 23 Positive in the table above. The data in the cell comprises one category.

The expected frequencies are what should have been. In many cases, you will be uncertain of the likely result and therefore the expected frequencies are averages based on the numbers in each category. If, for example, you have 100 observations in 2 categories, then the random expected frequencies would be 50 in one and 50 in the other. In our chart, with 4 categories, the expected frequencies would be 25, 25, 25 and 25. That is why we do not really want disproportionately small or large observation frequencies in a category unless we really mean it, as the statistical test will go all weak at the knees and say, "McGinty, we've struck gold".

On other occasions, the expected frequencies are not random, but are based on what we have come to expect from previous results. Let us say that the incidence of cheating has fairly predictable levels experienced by different sports. Then a new study takes into account a relatively novel phenomenon, perhaps the introduction of a very expensive (or very cheap) performance-enhancing drug. In such a case, our statistical testing will examine whether the proportions are fairly comparable or different from the previously recorded results. So if our previous knowledge, for example, indicates expected proportions of .25, .30 and .45 within three categories, we would be interested to find out if the newly observed frequencies in the three categories have similar or different proportions.

Quantification of categorical data is statistically simple but it does depend upon sound reasoning, taking into account logic within the research context.

The binomial test: a frequency test for dichotomies (either/or)

This test is for use with two categories only. Let us ask a sample of 30 individuals a simple question: "Do you think that titanium necklaces improve performance, yes or no?" 13 people said 'yes' and 17 people said 'no' (we're in the realm of fiction as usual).

If you are not using the **Frequencies.csv** file, just type in Yes onto your spreadsheet and copy it so that you have 13 of this category, and then do the same to get 17 negatives. In real life, it doesn't matter if they are in higgledy-piggledy order, as in the file. It also doesn't matter how you represent the levels. The first three columns from Frequencies.csv are the same for the purposes of the binomial test, be it Yes or No, the typical 1 and 2 of dummy coding, or my idiosyncratic 16 and 7. It is also possible to use summary data, which I will demonstrate shortly.

	Titanium1713Words	Titanium1713NumsA	Titanium1713NumsB
1	No	2	7
2	No	2	7
3	Yes	1	16
4	No	2	7
5	Yes	1	16
6	No	2	7
7	No	2	7

To use the test, press the Frequencies tab and select Binomial test from the drop-down table.

Proportion Test (2 Outcomes)

Titanium1713NumsA	**Titanium1713Words**
Titanium1713NumsB	
Titanium5050	
TitaniumSignificant	
Titanium	
Athletes	
AthleteFreq	
Nuisances	

☐ Values are counts

Test value 0.5

Additional Statistics

☐ Confidence intervals

Interval 95 %

Hypothesis

◉ ≠ Test value

◯ > Test value

◯ < Test value

> Bayesian Statistics

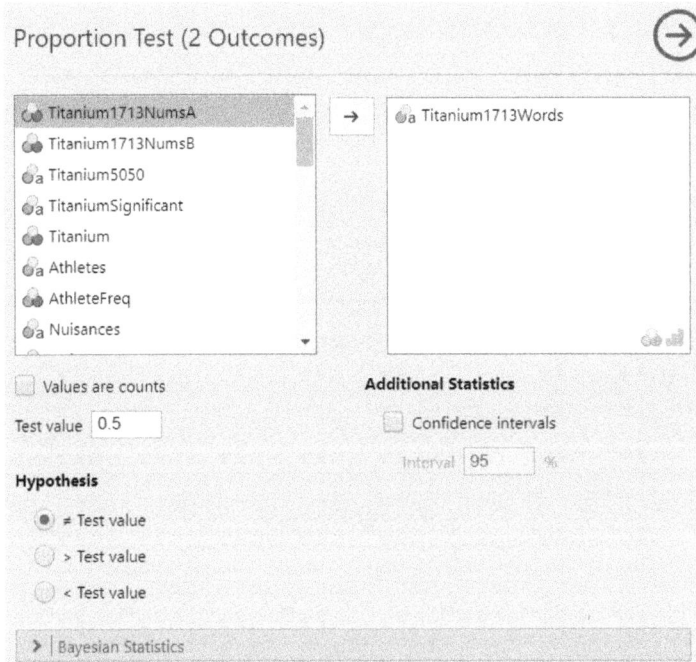

I have transferred the first variable to the box on the right (any of the three '17/13' variables will do the same thing).

Note that the Test value default is 0.5, which means a random, 50/50 expectation. As there are 30 observations, the expected numbers would be 15 for each level (Jamovi's term for category or condition). It assumes either that we have no assumptions about proportions or that the results of previous studies were actually 50/50. The value can be adjusted to anywhere between 0 and 1. If we chose 0.75, for example, then previous studies would have had a ratio of three-quarters for one of the two conditions, a quarter for the other.

Note also the Hypothesis option. The default, that the results are different from the Test value, is the test's equivalent of a two-tailed test. It is rare for researchers to use one-tailed categorical tests. If there were a comparison of different conditions both of which were expected to have some effect, then a one-tailed hypothesis may be justifiable.

Binomial Test

	Level	Count	Total	Proportion	p
Titanium1713Words	No	17	30	0.567	0.585
	Yes	13	30	0.433	0.585

Note. H, is proportion ≠ 0.5

As you can see, we get the counts of the level - it is always useful to check that your data entry is correct - the proportions (out of 1), and a high p value. There is support for the null hypothesis, that there is no clear difference from 50/50.

If you use the variable Titanium5050, you can see what the test does with a sample of exactly 50/50.

Binomial Test

	Level	Count	Total	Proportion	p
Titanium5050	No	15	30	0.500	1.000
	Yes	15	30	0.500	1.000

Note. H, is proportion ≠ 0.5

This shows equal proportions and the highest possible value of p.

Let's now look at a significant result. Use the TitaniumSignificant variable. Using the default settings, you will get this:

Binomial Test

	Level	Count	Total	Proportion	p
TitaniumSignificant	No	22	30	0.733	0.016
	Yes	8	30	0.267	0.016

Note. H, is proportion ≠ 0.5

We have a sample less keen on titanium. With a p value of 0.016, we can reject the null hypothesis of a similar number of responses on either side, unless of course we had specified a critical value of $p < .01$ or smaller. However, if we had good reason *before seeing the data* to expect a result in the 'No' direction, then we could select a one-tailed hypothesis (Test value); this gives a p value of 0.008, meeting the lower critical value.

But what if we had expected, from other studies, that our sample would be approximately 75/25? If you alter the Test value to .75 and click on the interface, you will then see a large p value against 'No'. In other words, we cannot reject the null hypothesis that No is similar to a three-quarters majority. Conversely, if you enter the value as .25, you will find the same p value against 'Yes', essentially a mirror image. So, the binomial test can test data against other proportions as well as 50/50.

Instead of raw data, you can use summary data. Type in the data like so (you don't even need to name the column heading).

🔵 Titanium
13
17

Then choose the 'Values are counts' option.

This gives the same result as with the raw data:

Binomial Test

	Level	Count	Total	Proportion	p
Titanium	1	13	30	0.433	0.585
	2	17	30	0.567	0.585

Note. H$_a$ is proportion ≠ 0.5

The multinomial test: a frequency test for more than two categories

The multinomial test may be considered as an extension of the binomial test. It covers more than two conditions and, like the binomial, can be used with even proportions as the default or can be set to compare the observed results with expected proportions. It is also known as the chi squared Goodness of Fit test; please do not confuse it with what is normally referred to as 'Chi squared', with which you may be familiar, which covers two variables in 2 × 2 and larger grids. That is the chi squared test of Association, which appears after this test.)

Throwing	RunningWalking	Jumping
80	28	42

Let us consider a categorization of athletes.

🐎a Athletes	👟 AthleteFreq
Throwing	80
RunningWalk...	28
Jumping	42

To use the test, press the Frequencies tab and select Chi Squared Goodness of fit from the drop-down table.

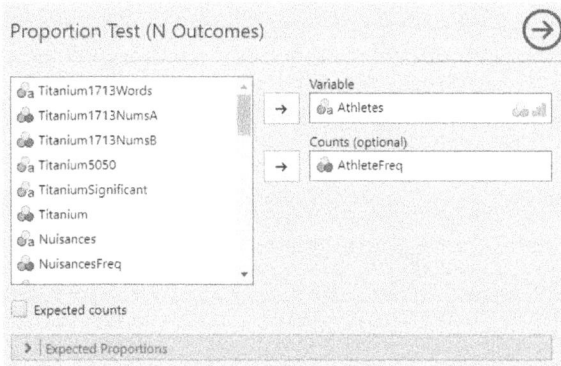

Proportion Test (N Outcomes)

The titles of the categories go into the Variable slot and the corresponding frequencies go into the Counts box.

Proportions - Athletes

Level	Count	Proportion
Jumping	42	0.280
RunningWalking	28	0.187
Throwing	80	0.533

χ^2 Goodness of Fit

χ^2	df	p
29.0	2	< .001

We have a large Chi squared statistic and the p value is a low one, below .001.

If you choose the Expected counts option, it will be clear that each level was 'expected' to be 50; there are three equal slices from our 'cake' of 150 responses. Our result tells us only that there is a significant difference from the expected, random, proportions. This does not say anything about any one category. If you wanted to isolate a specific condition, you would have to concatenate the categories and then run a binomial test for the remaining two categories, but the new categories would have to make sense. I am not sure about this example; I suppose you could separate the throwing 80 from the non-throwing 70.

For an example of a non-significant data set, let us say that we have asked schoolteachers to tell us about the worst problems for school sports: parental intrusion, coaches' over-emphasis on winning, and organizations demanding too much commitment.

Parents	Coaches	Organizations
35	37	33

From Frequencies.csv, transfer the Nuisances variable into the Variable slot and NuisancesFreq into the Counts slot. A p value of 0.892 is a clearly non-significant result. (The expected counts are 35 apiece; our observed frequencies, as in the table, do not vary much from these.)

As with the binomial test, this can be applied to proportions which are to be expected, often because of previous evidence.

Let us say that there are four different types of illegal drugs most used in a sport.

Drugs	DrugFreq
A	315
B	108
C	101
D	32

This data represents the proportions of illegal drugs according to the records.

We already have records of previous drug usage. Here, I open the Expected Proportions section and put in the percentages for each drug.

Level	Ratio	Proportion
A	56.2!	0.563
B	18.7!	0.188
C	18.7!	0.188
D	6.25	0.063

So, in the previous records, 56.25 of users took Drug A and so on. Do note that the percentages for the combined drug use should equal 100%. Another way I might have done this was to use a ratio. If I had divided the drugs into 16ths, I could have put in 9 for Drug A (9/16), 3 for Drug B, 3 for Drug C and 1 for Drug D, with the same results.

Proportions - Drugs

Level		Count	Proportion
A	Observed	315	0.5665
	Expected	312.8	0.5625
B	Observed	108	0.1942
	Expected	104.3	0.1875
C	Observed	101	0.1817
	Expected	104.3	0.1875
D	Observed	32	0.0576
	Expected	34.8	0.0625

χ^2 Goodness of Fit

χ^2	df	p
0.470	3	0.925

In the results, I show the 'Expected' statistics, which show a close correspondence to the recorded usage. This appears to reflect previous trends in drug usage. Returning to the technicalities, we see a small Chi squared statistic and a high p value, indicating a non-significant result: the observed frequencies are fairly similar to the expected frequencies.

The chi squared Test of Association: a frequency test for two variables

Often known as just Chi Squared, probably because most introductory textbooks teach only this categorical test, the chi squared Test of Association is used to find out whether or not there is a relationship between variables. The test is also known as the chi squared Test of Independence, perhaps a purer statistical definition, as the null hypothesis is that the variables will be independent of each other. We may want a 'significant' relationship; the computer seeks a lack of significance, the null hypothesis.

The contingency table allows us to study two or more variables in tandem. The statistical test itself, chi squared, distinguishes whether or not there is a relationship between the variables.

Before going into this in more detail, it is worthwhile knowing that there are two ways of entering the data. First, let us look at *raw data*. Here we use two variables, Ethnicity and Gender, each with two conditions/levels:

	Variable: Gender	
Variable: Ethnicity	Female	Male
Non-White	6	6
White	6	12

The image below demonstrates raw data entry from Frequencies.csv, showing some of the cases.

Ethnicity	Gender
Non-white	Male
Non-white	Male
Non-white	Female
Non-white	Female
White	Male
White	Male
White	Male
White	Male
White	Female
White	Female

For this test, press the Frequencies tab and select Independent Samples – Chi squared test of association – from the drop-down menu.

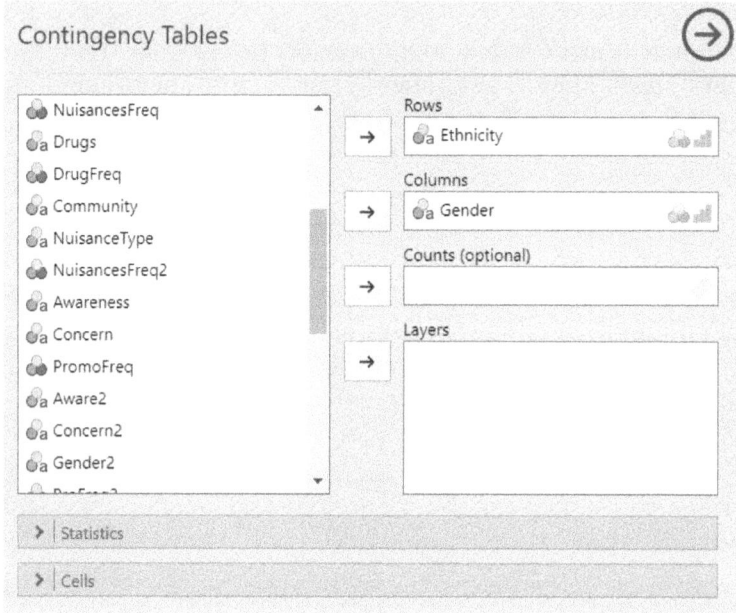

All you need to do is to transfer one variable into the Rows box and the other into the Columns box. This is only a 2 × 2 grid, White and Non-white versus Male and Female, so the order of rows and columns does not matter much. Where a variable has many levels, it may be a good idea to put the variable with more conditions into Rows for neater output.

The output shows a large p value, 0.361. We cannot reject the null hypothesis, that there is no significant relationship between the two variables Gender and Ethnicity.

Moving on, let us extend our interest in problems reported by sports teachers, as when looking at the chi squared Goodness of Fit test, by examining the situation in rural and urban communities. The chart shows the design, a grid of 2 × 3 (levels/conditions), although still only two variables, Nuisancetype and Community.

Variable: Nuisances	Variable: Communities	
	Rural	Urban
Parents	46	42
Coaches	23	17
Organizations	11	15

Note the exhaustiveness and exclusivity rule: We have 154 schools in our sample, all are allocated, each being allocated to just one of these cells.

Here we use *summary data*. The first two columns represent the two variables; between the two, we cover all of the permutations of levels across the two variables. There then follows the frequency for each cell.

Community	NuisanceType	NuisancesFreq2
Rural	Parents	46
Rural	Coaches	23
Rural	Organizations	11
Urban	Parents	42
Urban	Coaches	17
Urban	Organizations	15

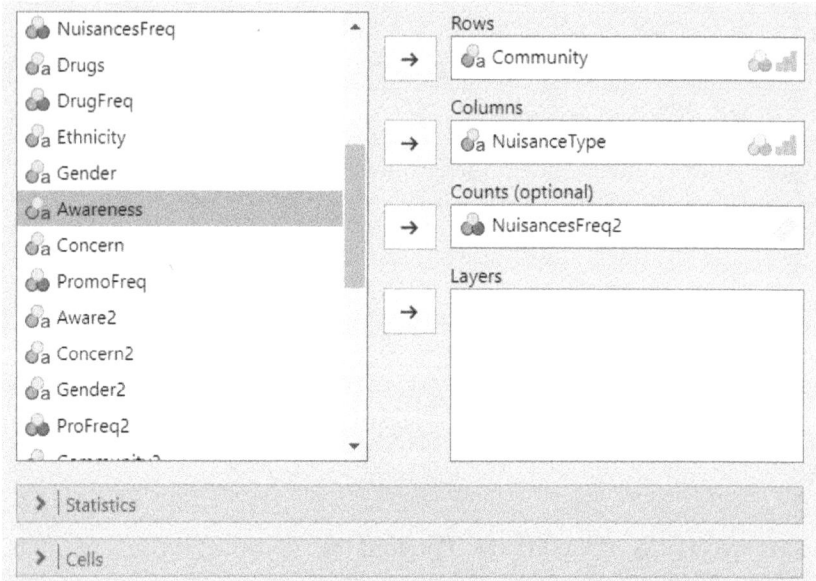

We transfer the two summary variables into the Rows and Columns boxes, placing the frequencies variable in the Counts box. In the output, you will see a p value of 0.481; the null hypothesis may not be rejected.

Give me a significant result, I think I hear you say.

We might be considering a promotional campaign, perhaps on engagement in exercise, or perhaps countering current myths about the role of energy drinks. Let us consider asking two questions: Have respondents seen or heard about an issue? Are they concerned about an issue? Do note that the test only allows us to consider whether or not there is a relationship between the two variables; it cannot ascertain causal direction.

		Concern	
		Serene	Worried
Awareness	Not Seen	30	12
	Seen	12	18

Awareness	Concern	PromoFreq
Seen	Worried	18
Seen	Serene	12
Not seen	Worried	12
Not seen	Serene	30

The data, as in the file, contains pairings of the different levels (2 × 2) across the two variables, Awareness and Concern. Each pairing has a frequency count, the number of observations per category. So we have 72 individuals, each allocated to just one cell.

The variables are placed in Rows, Columns and Counts, as in the previous example.

Contingency Tables

	Concern		
Awareness	Serene	Worried	Total
Not seen	30	12	42
Seen	12	18	30
Total	42	30	72

χ^2 Tests

	Value	df	p
χ^2	7.11	1	0.008
N	72		

We have a small p value, 0.008, which would meet the critical value of $p < .01$.

A note on samples: It is generally recommended that there should be at least 20 observations per sample, with at least 5 observations per cell. On the other hand, where you do have some cells with 5 or less (some say below 10), you may choose to go into the Statistics section and choose the Chi squared continuity correction (also known as the Yates

Correction). Actually designed for small numbers of expected cells, this adjustment gives a more conservative assessment, with a higher p value. However, it has a tendency to produce Type 2 errors (wrongly supporting the acceptance of the null hypothesis or, in layman's terms, falsely indicating non-significance). Many statisticians believe that it should never be used. It is certainly not necessary in this case.

To complicate your life further, you should know that very large samples often lead to very small p values. In fact, a large sample may appear to be 'more significant' than a small sample with the same relationship between observed and expected values. Here, measures of effect size, the magnitude of the relationships, really come into their own. They adjust for sample size and are thus more usable for larger samples than chi squared and the p value.

The most commonly reported of these 'measures of association', as they are also called, are Phi and Cramer's V, to be found in the Statistics section. Both have the useful attributes of running from 0 to 1, from no relationship to being exactly the same. Phi is usually read for 2 x 2 tables, with Cramer's V for larger contingency tables (for example, one variable is comprised of children, teenagers and adults, another of males, females and 'other gender').

In the promotional campaign example, a 2 x 2 table, both Phi and Cramer's V give 0.314 as the magnitude of the effect. Cohen (1988) provides a rule of thumb in which 0.1 is a small effect, 0.3 a medium effect and 0.5 a strong effect. His examples of their comparative magnitude are the difference in mean height between 15- and 16-year-old girls (small), the difference between 14- and 18-year-old girls (medium) and that between 13- and 18-year-old girls (large).

With larger contingency tables, such as a 3 x 3, smaller effect size statistics emerge. To interpret them, the df statistic displayed in the output next to Chi squared and the p value become of use. The degrees of freedom (df) refer to the number of factors free to vary in the calculations. The df statistic is calculated by taking the combined number of rows and columns and subtracting 1.

df	small	Medium	large
1	.10	.30	.50
2	.07	.21	.35
3	.06	.17	.29
4	.05	.15	.25
5	.04	.13	.22

This table is adapted from Cohen (1988). As can be seen, smaller Cramer's V statistics are accepted when larger grids are in use.

One question which may occur to you is what should you do when you come across figures such as .2 and .4? Personally, I would consider .4 a strong effect size; indeed, effect sizes of more than .5 are to be viewed with some suspicion, as they may indicate that the different variables are measuring similar concepts.

In general, however, it should be noted that the above recommendations are a rule of thumb. Knowledge of previous studies covering similar ground are generally considered to be more helpful – if you can find them. If in doubt, just cite the statistic without committing to a description of the size.

I have not consulted the ordinal tests, Gamma and Kendall's *tau-b*, as our variables are not gradable in terms of magnitude. A possible ordinal example would be responses to an online video about the use of steroids: one side of the grid would include indications that respondents watched the entire film, some of the film, or left at or near the beginning. The other side could comprise respondents with personal experience or an expert knowledge of steroid use, those with some understanding, and those with very limited knowledge. This assumes that we agree that both of these variables are gradable, in terms of levels of participation and of knowledge of the topic.

Gamma should be interpreted as follows: .75 to 1 = a Strong relationship; .5 to .74 = Moderate; .25 to .49 = Weak; < .25 = No relationship. If you look at Gamma in the (inappropriate) promotional campaign example, you will find that the result is just about within the moderate reporting category: essentially, tests for ordinal data, that is with categories of increasing magnitude, are generally more demanding than those merely seeking a relationship without such a direction. *Tau-b* may be better for grids with the same number of rows and columns in the table.

Remembering that the test result only comments on whether or not the overall relationship between the two variables is significant (or rather, allows us to reject the null hypothesis), it is often helpful to take a closer look at the grids. The Cells section provides a wealth of information about what has been going on inside the test. First, within Counts, select Expected to go alongside the default Observed counts.

Contingency Tables

Awareness		Concern		Total
		Serene	Worried	
Not seen	Observed	30	12	42
	Expected	24.5	17.5	42.0
Seen	Observed	12	18	30
	Expected	17.5	12.5	30.0
Total	Observed	42	30	72
	Expected	42	30	72

In this particular case, all of the cells have a strong difference between the observed count and the expected count. This is not always the case. In this particular instance, we can see that those who have seen the promotion are less likely to be serene and more likely to be worried than if the numbers were randomly distributed (the null hypothesis). The opposite is the case for those who have not seen the promotion.

Do not assume causality, however. It could be that seeing the promotion increases anxiety about the issue, but it is also possible that being anxious about an issue makes people more likely to look for materials which reflect their feelings. Another possibility is a mediating factor: perhaps additional news coverage could increase both anxiety and the tendency to follow the promotion.

In cases where things are not certain, it is possible to reduce categories in order to subject them to the chi squared Goodness of Fit test or the binomial. However, you would need to be sure that the concatenated categories make good theoretical sense or at least form a coherent rationale. The usual rules of exhaustiveness and exclusivity apply. Here, we strip away the Expected counts and just look at the default Observed counts.

Contingency Tables

Awareness	Concern		
	Serene	Worried	Total
Not seen	30	12	42
Seen	12	18	30
Total	42	30	72

In this case, you would use the outer figures to look at single variables. If examining Awareness, you would look to the far right, using the 42 and 30 for a binomial test. You would do likewise for the figures at the bottom for the Concern variable. With a larger grid than this, the proportions would be less simple and you would use the chi squared Goodness of Fit where a variable has more than two conditions.

Another thing that you can do is to study the Percentages within the Cells area. I find these particularly useful for reporting results. Personally, I would look at them one by one. If you have a particular area of interest, you may decide to view only one set of these.

Contingency Tables

Awareness		Concern		Total
		Serene	Worried	
Not seen	Observed	30	12	42
	% of total	41.7 %	16.7 %	58.3 %
Seen	Observed	12	18	30
	% of total	16.7 %	25.0 %	41.7 %
Total	Observed	42	30	72
	% of total	58.3 %	41.7 %	100.0 %

Total: The outer figures show the overall totals per variable. So if you wanted to report on the percentages for the Awareness variable you would look at the right and use Not Seen 58.3% and Seen 41.7% (58% and 42% respectively if rounding the figures). The same is true for the Concern variable, using the figures at the bottom. As before, these would vary more when you deal with a different-shaped grid. Inside, you have the possibility of reporting any of the categories in terms of percentages (for example, those who had seen the promotion and were worried comprised 25% of the sample). Often, it is a good idea to report percentages, as they usually demonstrate proportionality more effectively than absolute numbers.

Contingency Tables

Awareness		Concern		Total
		Serene	Worried	
Not seen	Observed	30	12	42
	% within row	71.4 %	28.6 %	100.0 %
Seen	Observed	12	18	30
	% within row	40.0 %	60.0 %	100.0 %
Total	Observed	42	30	72
	% within row	58.3 %	41.7 %	100.0 %

Chapter 7 – Categorical analyses

Row: This is where you are interested in the relationship between the figures *within the row variable*. To give this some flesh, imagine that we have replaced the 'Seen' row with Females and 'Not Seen' with Males. You would then be able to give figures for Females: "Of the females in the sample, a little over 70% were reasonably relaxed about the issue." For Males: "Of the males, 60% were worried."

Contingency Tables

Awareness		Concern		
		Serene	Worried	Total
Not seen	Observed	30	12	42
	% within column	71.4 %	40.0 %	58.3 %
Seen	Observed	12	18	30
	% within column	28.6 %	60.0 %	41.7 %
Total	Observed	42	30	72
	% within column	100.0 %	100.0 %	100.0 %

Column: Here, we view the proportions from the top down. Let us do a similar trick of the imagination by replacing 'Serene' with Urban and 'Worried' with 'Rural'. 71% of the urbanites have not seen the promotion. 60% of the country-dwellers have seen it. (As usual, we are still left with whether or not rural citizens were more sensitized to the issue and therefore more likely to see the promotion, or perhaps the promotion somehow appeared in places or media more likely to be seen by a particular social class, or some other mediating factor.)

A note on the Layers box. This is for additional variables. Here, we split the data we had on promotion between males and females:

☺a Aware2	☺a Concern2	☺a Gender2	☺☺ ProFreq2
seen	worried	female	10
seen	serene	female	7
not seen	worried	female	4
not seen	serene	female	16
seen	worried	male	8
seen	serene	male	5
not seen	worried	male	8
not seen	serene	male	14

If the 'Gender2' variable is placed in the Layers box, we will get the same overall results as previously, but also a separate examination of each of the gender levels. Beware of using more than one layer variable at any one time, as the results may be very difficult to interpret.

For charts, put the information into the log-linear regression dialog box, as discussed in the next section. Use the Estimated Marginal Means section.

Log-linear regression: modeling three or more categorical variables

If you have more than two categorical variables, log-linear regression starts with a large model representing the data and looks for more parsimonious smaller models. As the name indicates, this technique uses regression techniques, so if you haven't already read about regression, it might be sensible to do so in order to enhance your understanding of the output.

Let us consider a (fictional) study of performance-enhancing drugs and gender in three towns.

		Drug type		
Community	Gender	Diuretics	Steroids	Stimulants
Smallville	Male	38	16	26
	Female	14	32	24
Littleville	Male	46	11	24
	Female	8	34	27
Tinyville	Male	63	13	12
	Female	19	27	16

From our previous research, we expect Tinyville to be most involved in diuretics, the citizens of Smallville to be, on average, the least involved in diuretics and Littleville to come somewhere in between. We also have reason to believe that males are more likely to be involved in taking diuretics than females. What we don't know is whether or not there will be an interaction between communities and gender involvement in drug-taking.

Communi...	DrugType	Gender3	CrimeFreq3
Smallville	Diuretics	Male	38
Smallville	Diuretics	Female	14
Smallville	Steroids	Male	16
Smallville	Steroids	Female	32
Smallville	Stimulants	Male	26
Smallville	Stimulants	Female	24
Littleville	Diuretics	Male	46
Littleville	Diuretics	Female	8
Littleville	Steroids	Male	11
Littleville	Steroids	Female	34
Littleville	Stimulants	Male	24
Littleville	Stimulants	Female	27
Tinyville	Diuretics	Male	63
Tinyville	Diuretics	Female	19
Tinyville	Steroids	Male	13
Tinyville	Steroids	Female	27
Tinyville	Stimulants	Male	12
Tinyville	Stimulants	Female	16

Each permutation of levels has its own row, including a count cell.

From the menu, press the Frequencies tab and select Log-Linear Regression. As well as transferring the variables as usual, I have also opened the section referred to as Reference Levels. Each variable has a reference level, the baseline level (condition) with which other levels are compared. The default baseline is set by alphabetical order. In this example, the community variable has three levels, Littleville, Smallville and Tinyville. With Littlefield as default, we would get pairings of Smallville-Littlefield and Tinyville-Littlefield. In this example, I prefer to have Tinyville as the baseline, with the pairings Littleville-Tinyville and Smallville-Tinyville; so Tinyville is in the level slot. Male is also baseline (and Diuretics by default).

Model Coefficients

Predictor	Estimate	SE	Z	p
Intercept	4.143	0.126	32.885	< .001
Community2:				
Littleville – Tinyville	−0.314	0.194	−1.622	0.105
Smallville – Tinyville	−0.506	0.205	−2.461	0.014
DrugType:				
Steroids – Diuretics	−1.578	0.305	−5.181	< .001
Stimulants – Diuretics	−1.658	0.315	−5.265	< .001
Gender3:				
Female – Male	−1.199	0.262	−4.580	< .001
Community2 ✳ DrugType:				
(Littleville – Tinyville) ✳ (Steroids – Diuretics)	0.147	0.453	0.325	0.745
(Smallville – Tinyville) ✳ (Steroids – Diuretics)	0.713	0.426	1.674	0.094
(Littleville – Tinyville) ✳ (Stimulants – Diuretics)	1.008	0.403	2.499	0.012
(Smallville – Tinyville) ✳ (Stimulants – Diuretics)	1.279	0.405	3.158	0.002
Community2 ✳ Gender3:				
(Littleville – Tinyville) ✳ (Female – Male)	−0.551	0.464	−1.187	0.235
(Smallville – Tinyville) ✳ (Female – Male)	0.200	0.408	0.491	0.623
DrugType ✳ Gender3:				
(Steroids – Diuretics) ✳ (Female – Male)	1.930	0.427	4.517	< .001
(Stimulants – Diuretics) ✳ (Female – Male)	1.486	0.463	3.211	0.001
Community2 ✳ DrugType ✳ Gender3:				
(Littleville – Tinyville) ✳ (Steroids – Diuretics) ✳ (Female – Male)	0.948	0.670	1.414	0.157
(Smallville – Tinyville) ✳ (Steroids – Diuretics) ✳ (Female – Male)	−0.238	0.612	−0.389	0.697
(Littleville – Tinyville) ✳ (Stimulants – Diuretics) ✳ (Female – Male)	0.381	0.663	0.574	0.566
(Smallville – Tinyville) ✳ (Stimulants – Diuretics) ✳ (Female – Male)	−0.568	0.626	−0.907	0.365

The chart gives us essential information about the variables and the relationships between them. The statistics are the coefficients ('Estimate'), standard error, the z values, which are measures of deviation away from the mean, and the p values. (You will need Jamovi on full screen to see all of this at once.)

Using the p values, we can see that the test supports the null hypothesis for the relationship between community and gender; this and, unsurprisingly, the higher level model of an interaction between community, drug type and gender, can be discounted. The relationships between drug type and gender and between community and drug type may be worthy of further analysis.

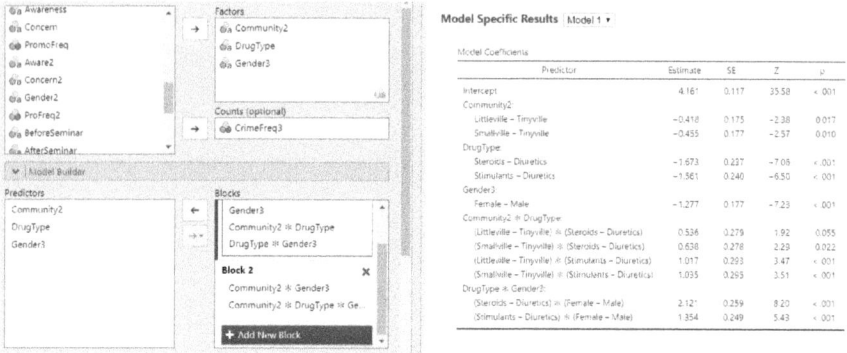

To get a more parsimonious model with rather altered statistics, create Block 2 in the Model Builder section and transfer the less useful interactions into it. So in Model 2, first place the three-way interaction and then the relationship between community and gender. Toggle to Model 1, which is created by Block 1 in the Model Builder.

You can also look at the measures of fit, as found in the Model Fit section. Some information about AIC and BIC can be found in Chapter 6, on the subject of hierarchical regression.

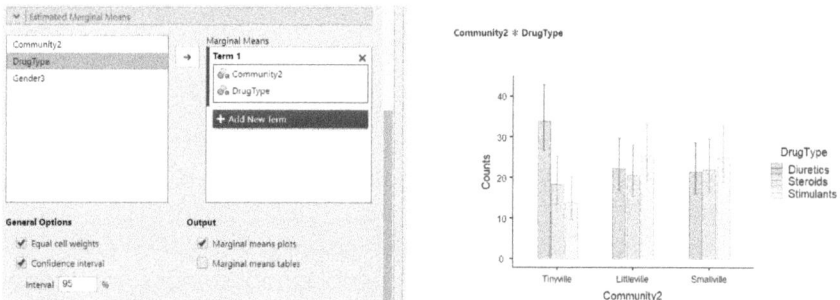

Go into the Estimated Marginal Means section. Where you would like to see an interaction, put the relevant variables into a 'Term' box. Note that the order in which you enter the variables affects the way the graph is composed.

Where next? You can go to Chapter 14, on logistic regression, to learn about some of the other regression statistics. Alternatively, collapse the study into interactions of two effects, using the chi squared test of association, in the previous section of this chapter. Further analyses could include the 'observed' and 'expected' observations and percentages.

The McNemar test: correlated dichotomies (linked pairs)

Introduction

The main point about this test is that it examines changes in paired attributes. As with the rest of the tests shown here, the attributes are nominal: yes and no, for and against, solved or not solved, and so on.

The test is not just at the end of the chapter because it is rather different from the other tests, but also because of its level of difficulty. People sometimes have trouble understanding the function of the test, or its implementation. To make this clearer, and to get an idea of its breadth of potential application, we will first look at a few possible examples. In each case, we consider matching pairs.

Examples

After a trial of a controversial training technique, do players of a sport change their views, favorable or unfavorable, on the technique's effectiveness? Data collection consists of counting the number of responses in four categories: before/favorable, before/unfavorable, after/favorable and after/unfavorable.

Another example could ask referees, before and after a national change in how rules are enforced, whether or not they believe the reform is worthwhile. Another could be that some children with behavioral problems are given a course of boxing lessons; before and after the course, are they involved in gang conflict or not?

Design

To get the test to work properly, you need to have a clear outline of your study, set out in a particular way, which is then translated into suitable data entry for Jamovi.

The study's figures must be read across the upper row from left to right and then the lower row left to right. The top left should be a double positive (for example: before yes, after yes), with the bottom right being a double negative (for example: before no, after no). The other diagonal are mixed results (for example: before yes, after no, or vice-versa).

1+ +	2 + -
3 - +	4 - -

People (all right, me) sometimes have problems visualizing what is going on, so I will show two more tables to illustrate how the information should be tabulated before data entry..

	After	
Before	Yes	No
Yes	A	B
No	C	D

As you might expect, A is followed by B, is followed by C and then by D. *A = Before, Yes and After, Yes; B = Before Yes, After No; C = Before No, After Yes; D = Before No, After No.*

Before Yes & After Yes	Before Yes & After No
A	B
Before No & After Yes	Before No & After No
C	D

Here, the same logic is shown with a slightly different format. Use whichever you find easiest to use, as long as the logic is the same. Another way of explaining the logic is to say that consistent results are in the top left to bottom right diagonal (A and D), with altered results in the opposite diagonal (B and C).

Here I provide adaptations of tables from Fay (2015):

	Experimental	
Control	Fail	Pass
Fail	A	B
Pass	C	D

	Boxing course taken	
Gang conflict	Yes	No
Yes	A	B
No	C	D

In these last two examples, time – as in 'before and after' – is not the issue. In the first table, the connection is whether or not the same sports test has been passed or failed by members of two different groups (experimental or control, but could be two age groups, etc). In the second case, boxing course participation and gang conflict are juxtaposed in the same 'plus and minus' grid. In all cases, we have correlated pairs of categories; they must of course intersect.

Applying the test

Let us now use some actual figures and try out the test.

Bodybuilders found to have used steroids are offered a seminar on health repercussions. In particular, they learn about the threat to their reproductive systems.

Seminar on the dangers of taking steroids			
		After	
		Concerned about health risks	Sceptical about health risks
Before	Concerned about health risks	20	2
	Sceptical about health risk	12	16

This chart is typical of the 'before and after' design for this test, the top left figure is a 'double positive': some participants were worried about steroids' threat to health both before and after the seminar. The bottom right is a 'double negative': they were unconcerned at the start and persisted in that view after the seminar. The top right figure is a counter-intuitive 'plus-minus': 2 participants who had been concerned are now less worried. The bottom left is a 'minus-plus': once they realized that steroid usage was a serious threat to their health, 12 former sceptics became concerned. (As usual, this is a fictional study.)

Use the Frequencies.csv file.

🔵ₐ BeforeSeminar	🔵ₐ AfterSeminar	🔵 AttitudeCount
Concerned	Concerned	20
Concerned	Sceptical	2
Sceptical	Concerned	12
Sceptical	Sceptical	16

To start with, I advise adhering to this data entry format, exactly as it stands: double positive on row 1, positive and negative on row 2, negative and positive on row 3 and double negative on row 4.

Now press the Frequencies tab and select Paired Samples - McNemar test.

Paired Samples Contingency Tables →

			Rows	
👁 Concern...				
👁a Gender2	→		👁a BeforeSeminar	
👁👁 ProFreq2			Columns	
👁a Community2	→		👁a AfterSeminar	
👁a DrugType			Counts (optional)	
👁a Gender3	→		👁👁 AttitudeCount	
👁👁 CrimeFreq3				
👁a BeforeTraining				
👁a AfterTraining				

☑ χ²   **Percentages**
☑ χ² continuity correction   ☐ Row
☐ Log odds ratio exact   ☐ Column

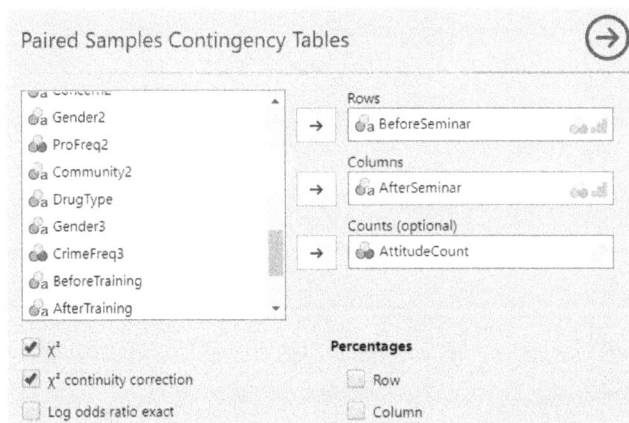

Our null hypothesis is that there is no appreciable change in attitudes to steroids as to whether or not they constitute a threat to reproductive health. I have chosen both Chi squared and the correction, preferring both to agree.

Contingency Tables

	AfterSeminar		
BeforeSeminar	Concerned	Sceptical	Total
Concerned	20	2	22
Sceptical	12	16	28
Total	32	18	50

McNemar Test

	Value	df	p
χ²	7.14	1	0.008
χ² continuity correction	5.79	1	0.016
N	50		

In this case, there is little controversy. The evidence supports rejection of the null hypothesis. There would appear to be a significant difference. If you go back to the table with the results, it seems that a substantive number of participants (12) who have previously not seen steroids as damaging to their health have changed their minds after the seminar.

Fitness training sessions			
		After training	
		Up to standard	Below standard
Before training	Up to standard	7	4
	Below standard	9	5

The next example also comprises a 'before and after' context, but shows a non-significant result. Before and after a pre-season series of training sessions, rugby union players are tested to see whether or not they have achieved acceptable levels of fitness.

BeforeTraining	AfterTraining	StandardCount
Standard	Standard	7
Standard	Below standard	4
Below standard	Standard	9
Below standard	Below standard	5

Contingency Tables

	AfterTraining		
BeforeTraining	Below standard	Standard	Total
Below standard	5	9	14
Standard	4	7	11
Total	9	16	25

McNemar Test

	Value	df	p
χ^2	1.92	1	0.166
χ^2 continuity correction	1.23	1	0.267
N	25		

Chapter 7 – Categorical analyses

The result is clearly insignificant. The null hypothesis, that there are no appreciable differences in the proportions of acceptable fitness levels after the course, cannot be reasonably rejected. *

	Boxing course undertaken	
Gang conflict	Yes	No
Yes	21	9
No	2	12

Children with behavioral disorders have been offered a course in boxing. One of the desired outcomes is a reduction in gang conflict. The records are checked and we look at whether or not each child attended a course and whether or not they were involved in gang conflict. With thanks to Michael Fay (2015) for the (adapted) data set.

GangConflict	BoxingCourseTaken	ChildrenCount
Yes	Yes	21
Yes	No	9
No	Yes	2
No	No	12

---

*The null hypothesis in technical terms is that there is no appreciable change in the marginal probabilities. Without going into the details too far, a marginal probability is that of an independent event happening. If you mess with the numbers, you will find that the double positives and double negatives are of no importance. In the meantime, the two 'mixed' values – where the results have changed in either direction – can be interchangeable; these two values are examined by the test, to see if there is much of a difference in direction.

Contingency Tables

	BoxingCourseTaken		
GangConflict	No	Yes	Total
No	12	2	14
Yes	9	21	30
Total	21	23	44

McNemar Test

	Value	df	p
χ^2	4.45	1	0.035
χ^2 continuity correction	3.27	1	0.070
N	44		

If we use Fisher's critical value of $p < .05$, then we are rather uncertain. The sample is a small one and therefore the uncorrected chi squared statistic is a little suspect. However, Yate's correction, although traditionally recommended in such a case, has a reputation for being overly strict. The statistic does not look prohibitively large. Fay (2015) suggests using the *Exact McNemar* test as an arbitrator in such a situation.

Select the Log odds ratio exact option.

McNemar Test

	Value	df	p
χ^2	4.45	1	0.035
χ^2 continuity correction	3.27	1	0.070
Log odds ratio exact	−1.50		0.065
N	44		

According to the Fisher criterion, we should not reject the null hypothesis.

Number of variables	Test
1 - two conditions	Binomial test
1 – more than two conditions	chi squared Goodness of Fit, aka multinomial test
2 - plus possible layers	Contingency tables, aka Chi squared Test of Association, aka 'Chi Squared'
3 or more	Log-linear regression
2 - linked pairs	McNemar test

Chapter 8 – Exercises

These exercises allow you to test your learning of Chapters 5, 6 and 7.

Questions

Question 1

Members of a focus group of referees decide between two strategies for dealing with disciplinary problems (no 'abstaining'). Of 40 opinions offered, one of the strategies was chosen by 26 people. What test should be used and what is the outcome? Accept a significance level of $p < .05$

Question 2

On a rating scale of difficulty, are male sports students more likely than females to see the dance module as stressful? Students of both genders were given the same five point rating scale to complete. What test is appropriate?

Question 3

Assume that educational background has been shown to influence attitudes to rugby. You have a range of variables such as the level of parental education, age of children and family income. What method would be most appropriate?

Question 4

In a sample of further education colleges, a study is being undertaken of the relationship between gender and management levels: principals, SMT (senior management team), HOD (heads of department) and senior lecturers.

	Principals	SMT	HOD	Senior Lecturers
Female	6	11	20	30
Male	8	18	16	32

What method should be used? Are there significant differences?

Question 5

Supporters of different football (soccer) teams are asked which of three explanations they prefer for why teams succeed in winning league titles.

Belief A = money; Belief B = management; Belief C = supporters.

At Beceister City, 100 favored Belief A, 230 favored Belief B and 150 favored Belief C. Banchester United favored Beliefs A, B and C as follows: 600, 400, 100. Belsea favored Beliefs A, B and C as follows: 180, 120, 100.

Consider an appropriate design and find out if the test results are significant.

Question 6

A correlation matrix containing a lot of variables includes many correlations at .9 What should you do?

Question 7

Can a correlation coefficient of .2 be significant?

Answers

Answer 1

This 'yes or no' situation, a dichotomy, can be subjected to the binomial test.

If we had a clear and well-reasoned rationale about one strategy being superior to the other, prior to examining the data, then a one-tailed hypothesis could have been used: the binomial test would have supported rejection of the null hypothesis. Unless this was the case, the two-tailed hypothesis needed to be chosen: the result would not be significantly different from chance.

Answer 2

Mann-Whitney examines the differences between two different sets of individuals. (If the rating scales had been calibrated, the independent samples ('unpaired') t test could have been used.)

Answer 3

Multiple regression. You would want to see how well different variables fitted into a model.

Answer 4

Chi squared Test of Association. The result is non-significant.

If you collated the different status categories, ignoring gender by adding the two groups together, and subjected them to the multinomial (Goodness of Fit) test, you would see a clearly significant result.

You could also use the binomial test to study the two genders, ignoring status. As the number of men and women in this sample are quite similar, this would be a non-significant result.

Answer 5

	money	manager	supporters
Beicester	100	230	150
Banchester Utd	600	400	100
Belsea	180	120	100

Chi squared Test of Association is the appropriate test, with a clearly significant result.

If you look at the actual and expected counts, you will see that Banchester United supporters are much more likely to believe in the influence of money on performance and less likely to see supporters to be an important factor. Beicester City supporters are in particular less likely to see money as influential.

Answer 6

The problem is likely to be collinearity. Use the test available within the linear regression assumption checks. Some of your variables probably have very similar meanings and should be removed. Similarities are useful, but not duplication. When your cull has lowered the collinearity of your data set, you could then use principal components analysis (see Chapter 13) to reduce data further, allowing a more in–depth investigation of the correlations.

Answer 7

This sort of correlation coefficient can be quite common when dealing with large data sets and may be accompanied by an acceptable p value. If you are interested in the magnitude of the effect, perhaps for applied usage, you may wish to consider the effect size, .04. In some contexts, 4% of the variance may be important; in others, this would be negligible.

Chapter 9 – Reporting research

As each university, and often each department, has its own guidelines for reporting, students are urged to consult the relevant guidelines. The basics, however, are the test statistic, the p value and, using parametric tests, the degrees of freedom (calculating variable and case numbers).

Although primarily about applied research, the following tips also work well in academia. Three issues are of central importance: The type of data, the target audience and the type of graphs being shown.

Data – absolute or averages?

Actually, you also need to consider this during your analysis. Even experienced analysts concern themselves about what is appropriate. Textbooks always make it look easy, but time after time, you will need to consider that apparently simple question, do I use the actual numbers or do I use a measure of central tendency (mean, median or mode)?

As each study is different, I doubt if there is a simple answer, but let us look at a few examples. If we wish to compare government expenditure on sport in different countries, then absolute numbers *probably* make more sense. On the other hand, if we are interested in the lives of individuals in those countries, for example, their expenditure on sports, gym memberships and consumption of energy drinks, then some type of average or indexing (collecting together 'baskets' of information and using a single scale), is probably more appropriate.

Another issue is the size. Let's start with small numbers. If you have been interviewing people or running focus groups, and there were only 10 people, would it really be appropriate to say "70% of respondents believed in the efficacy of this policy" when in fact those respondents were just a magnificent 7? I think I'd go for the absolute numbers.

Moving to larger numbers, think about government expenditure. Does the average person really know if 10 million dollars is a large or small amount by the standards of the day? Also, if comparing the expenditure with other countries, with different sizes of populations and different currencies, would not averages be more helpful? In many cases, one uses both sets of figures, but it helps to know which to emphasize in order for the audience to follow the logic of your study.

A few general points may help, but they are not eternal verities. For example, I may say "use the median" on certain occasions, but your employers may demand the mean at all times. Here are some general suggestions:

- Make it clear what type of data is being used.

- If using classification data (nominal/categorical/qualitative), you will typically show absolute numbers, although they may some-times be accompanied by the proportions.

- If you use a measure of central tendency, do make it clear which you are using.

- Use the mode to represent the most common response.

- Use the median to represent 'lumpy' data, the sort with which we usually use non-parametric tests.

- Use the median where continuous data is skewed away from the normal distribution.

- Use the mean for the results from parametric tests.

It is fairly common practice to accompany your main figure with the standard deviation (SD), a standardized way of representing the dispersion from the mean, positively and negatively. This may be useful for experienced researchers, but I still think that the median is effective for skewed data, as it is insensitive to extremes in the data.

Different audiences

Try to consider the likely level of sophistication of your audience. Here I mean their statistical understanding, although their likely views on the topic may also be of relevance!

The reader or member of the audience could be the commissioners of the research, or a line manager, or members of the public. To some extent you will have to guess, but I suggest three rough levels of statistical understanding.

The sophisticated audience is likely to know at least as much about statistics as you do and probably a lot more. The intermediate audience may remember vaguely about significance levels (usually the critical value of $p < .05$). The unsophisticated audience will not understand the difference between the word significance in its statistical sense and its dictionary usage; p values will be meaningless.

Let's look at some of the different concepts and consider the likely audiences:

- The null hypothesis – in general, I would not use this outside of academia. 'Significant' and 'non-significant' will normally do.

- p values – 'p = 0.042' is only for the most sophisticated audiences.

- Critical values – '$p < .05$' is ok for sophisticated audiences. Intermediate audiences may need a brief introduction on the occasion of its first usage, that it represents the probability of the effect being chance, lower figures suggesting that flukes are less likely and that being smaller than .05 indicates likely statistical significance.

- One-tailed and two-tailed hypotheses/tests – only sophisticated audiences will want to know about this and even for such an audience, a discussion of the expected direction would be helpful.

- Effect sizes – terms such as variance are only for sophisticated audiences, similarly r squared and other such statistics. 'Large', 'medium-sized' and 'small' effects would be suitable for an intermediate audience. Be even more sparing with unsophisticated audiences: mentions of particularly large and small effects will suffice, where they are of importance.

- Tests – sophisticated audiences will want to know which were used. Occasional usage may help intermediate audiences to believe that you know what you're doing, for example, "the result was significant to $p < .05$ (Mann-Whitney)". I would be inclined to omit this altogether when presenting to a lay audience.

- How you organized your data – in general, non-university audiences do not want to know about this, unless it is strictly of relevance to the study as a whole or you happen to think it important for a specific audience (maybe research commissioners, who paid for the work). You should always keep records, of course. In the case of your omitting outliers, a sophisticated audience should be informed, as they are likely to understand the relevance or otherwise of the data.

In general terms, the lay reader wants relevant results, ones with a bearing on the purposes of the research. It should not represent an academic thesis and while not an entertainment, should be readable and cogent.

Ah, says the worried, what happens if my audience is of *mixed levels of sophistication?* If you are fairly sure that your audience is highly variegated, and you believe that it is important that the lower level reader is not made to feel like a lemon (or whichever is your least favored fruit), then stratify your report. Perhaps put your most basic comment as the major part of a slide, followed by a more sophisticated comment in parentheses (e.g., "There was a significant difference between groups of self-employed respondents and those who were employed ($p < .05$ two-tailed).") If you know that you have even more fanatical stats-hounds in the audience, then consider footnotes.

Graphics

This is purely my own take on this. I try to limit my output on descriptive statistics to charts showing columns, charts showing horizontal bars and pie charts (leaving aside specialist charts such as those for correlations, factorial ANOVA and Kaplan-Meier plots for survival analysis). In general, I prefer columns for contrasting data, but bars come in useful if you have a lot of variables and/or lengthy titles, so that you need to spread down the page. I prefer to keep it simple, with one set of bars representing one effect, rather than layers or multiple meanings. Complex charts may mean something to you, because you've had your head in the data for extended periods; they may mean a lot less to your audience.

Pie charts are for exclusive data, where all the proportions are accounted for. If contrasting different groups, I generally prefer to have multiple pie charts, each representing a different group, rather than doughnuts and other concoctions.

Spreadsheets can also create simple correlation graphs (scatter plots), including a facility for adding a trendline. Other, quite complex graphs are also available to handle a variety of situations. If you choose to show one with layers, for example, be prepared to explain what it represents.

In general, I prefer to use graphics from a spreadsheet program rather than a specialist package. They are easier to tweak, adding titles, playing with the fonts and dealing with scales.

To a live audience!

Content needs to be limited – like short reports, only more so.

There are two particular things that will bore your audience and thus detract from your message: too much talking and information overload. Continuous talk is conducive to sleep or at best, lack of attention to what you think is important. The following practices are suggested for avoiding information overload and boredom.

- Try showing only one idea at a time, with a graph and the relevant statistics (depending on the audience) on one page.

- Try to keep slides interesting. Maybe have information sliding in from the side (not that I've ever mastered this). Avoid large slabs of text.

- Too much color can be distracting. Black and white is more effective than you might think.

While written reports require well-formed grammatical sentences, a presentation slide does not require all of the usual connecting phrases (although the voice-over does). The screen itself can have things like:

- "significant difference between the three body types"

- "moderate effect size"

- "limitations in available data"

- "implications for research into warm-up techniques"

You then provide a commentary as you read from the screen. "We found a significant difference…" – including things of interest. "The lack of data pertaining to injuries derived from this particular training technique raises the question of how far this data can be generalized. Further research may be worthwhile in this area."

Just repeating what you have written on the overhead chart is a terrible practice: your average member of the audience will wonder why they have turned up to listen. It is a good idea to prepare additional comments. At least do it in your head, but I think that rehearsing a couple of times is usually more effective.

For the purposes of smooth presentation and the avoidance of stage-fright, relatively inexperienced presenters may find it helpful to do a rehearsal in front of sympathetic colleagues. Be professional, addressing your audience at rehearsals as if they were your formal audience. This increases the chances of your being able to move into automatic mode when you are doing the real thing, unselfconsciously saying what you want to say.

You don't need to learn your words by heart. The screenshots are your prompts. Remember the few additional things you want to add; mentally associating them with the key phrases should help things to run smoothly. (And if it is your first time, don't dwell on the fact. Most of your audience will have done the same thing or will have to do so in the near future.)

In general, I would say that the key to successful reporting, oral or written, is effective categorization. Categorize, put the categories into a meaningful order and omit those which are likely to confuse or to bore unnecessarily. If some tedious things must be retained, then put them in a place where they are accessible but not center stage.

Part 3
ANOVA extended

Chapter 10 – Factorial ANOVA and multiple comparisons

Factorial ANOVA deals with more than one effect, or 'factor', at a time. As with the one-way ANOVA, we can examine an effect and, with multiple comparison tests, its constituent conditions ('levels'). Where the one-way ANOVA considers only differences, factorial ANOVA also allows us to look at relationships between factors, called interactions. That we can examine both differences and relationships at the same time is because of the underlying statistical model, the General Linear Model (GLM). GLM underlies not only factorial ANOVA but also one-way ANOVA, t tests and regression.

Factorial ANOVA is generally used for two-way or three-way analyses. It can be applied to more than three factors at a time, but additional factors make it hard to interpret the results. [*]

---

[*]It is also possible, using the same logic in reverse, to apply ANOVA to only two conditions within a single variable, but the t test gives us sufficient information.

Typical case studies

Factorial analysis of variance – within subjects

The same group of sprinters tests the effects of three different diets. Each diet period is subdivided into periods using different types of energy-enhancing drink.

Factorial analysis of variance – between subjects

Are schoolchildren in ability groupings more likely to perform well at swimming lessons compared to those in mixed ability groupings? Does gender have a part to play?

Factorial analysis of variance – mixed design

Sprinters are trying out different diets at different times. The results are also analyzed according to gender.

Effect sizes for factorial ANOVA

Clark-Carter (1997) makes the following recommendations: 0.14 is large, 0.06 is medium and 0.01 is small. This is extrapolated into bandings by Kinnear and Gray (2004):

Large: $> .1$ (more than 10% of the variance)
Medium: 0.01 to 0.1   (1% to 10% of the variance)
Small: $< .01$ (less than 1% of the variance)

Repeated Measures Two-Way ANOVA

We wish to review the effectiveness of a training campaign against sexism in sport. The data represents the number of recorded incidents within ten different sports. One of the factors is the intervention itself, divided into phases: before, during and after the campaign. The other factor is time, to see if differing moods in mornings and afternoons may affect responses.

For this exercise you will need to open the TwoWayRepeatANOVA.csv file.

Case	PreCampAM	PreCampPM	CampaignAM	Campaign...	PostCampAM	PostCampPM
1	6	8	4	5	6	8
2	4	5	3	4	4	4
3	9	8	6	5	8	6
4	7	4	7	6	8	8
5	6	7	6	6	7	6
6	7	8	5	7	6	7
7	5	5	4	3	4	5
8	6	8	4	5	5	7
9	4	3	3	4	5	6
10	6	9	4	6	5	8

The data set looks similar to a One-Way Repeated Measures ANOVA, but each column in fact contains subdivisions of the data:

Case	Pre-campaign (1)		Campaign (2)		Post-campaign (3)	
	AM (1)	PM (2)	AM (1)	PM (2)	AM (1)	PM (2)
1	6	8	4	5	6	8
2	4	5	3	4	4	4

The order is Factor 1 (1) with Factor 2 (1), then Factor 1 (1) with Factor 2 (1) and so on.

Press the ANOVA tab and select Repeated Measures ANOVA from the drop-down menu. At first, this is set up like the Repeated Measures one-way ANOVA.

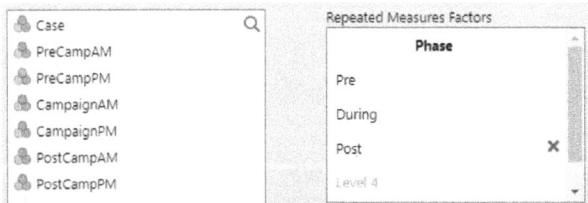

In the Repeated Measures Factors box, we replace RM Factor 1 with the name of the first factor (here, Phase) then adding in the conditions (Pre, During, Post) directly underneath the factor name.

Now, we do the same thing underneath for RM Factor 2 (using Time) and then its conditions/levels.

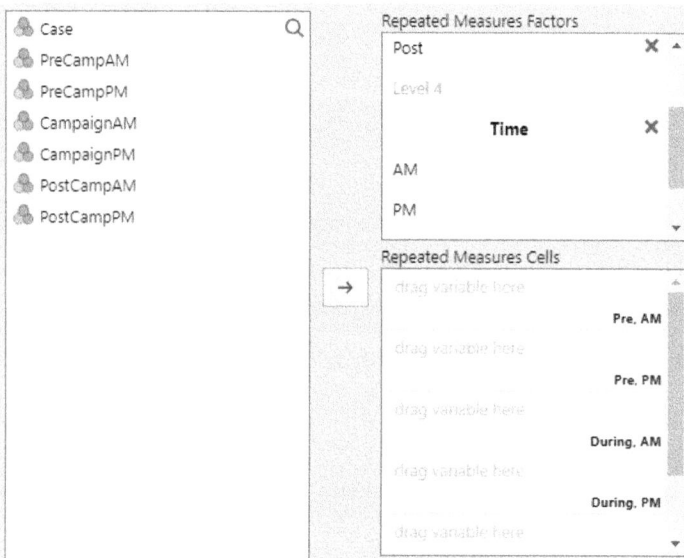

We can now see in the Repeated Measures Cells that permutations of the levels of the two factors have been automatically inserted on the right-hand side. The first level of the first factor has been conjoined with the first level of the second factor; then the first level of the first factor is matched with the second level of the second factor; and so on.

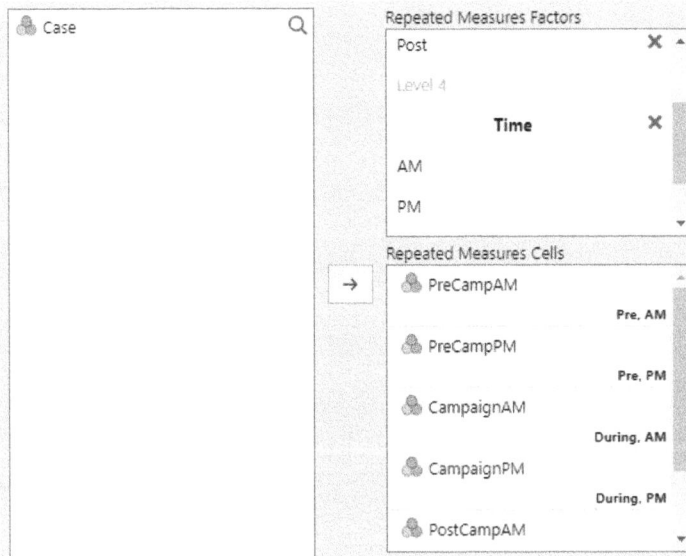

You then need to transfer the variables on the far left to the left-hand side of the Repeated Measures Cells. This example is straightforward, but as with other variants of factorial ANOVA, you need to be careful to ensure that each entry matches its logical pairing on the right.

Before analyzing the results, you should go to the Assumption Checks section and select Sphericity tests. These apply when there are more than two levels in a variable; in this instance, there are no significant results. If your data fail to meet this assumption, which relates to variance pertaining to pairs of groups and can lead to Type 1 errors (mistakenly assuming significance), a different reading of the results is required. This is covered in the worked example on the two-way mixed ANOVA later in the chapter.

Within Subjects Effects

	Sum of Squares	df	Mean Square	F	p
Phase	24.400	2	12.2000	9.438	0.002
Residual	23.267	18	1.2926		
Time	4.817	1	4.8167	1.883	0.203
Residual	23.017	9	2.5574		
Phase * Time	0.133	2	0.0667	0.159	0.854
Residual	7.533	18	0.4185		

The first factor, Phase, has a large F ratio and a p value of 0.002. The second factor is not significant – the time of day is clearly immaterial – and the same can be said for the interaction between the two factors (Phase * Time). If you choose an effect size (I usually use Partial Eta Squared), you will see on the ANOVA table that the effect size for Phase is a very large .512, over 50% of the variance.

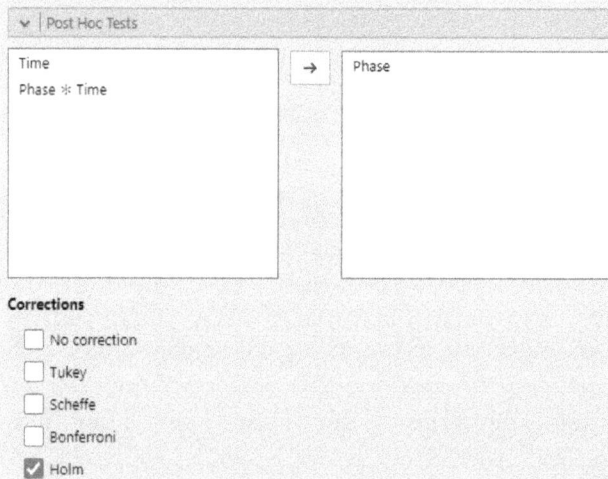

As the Phase main effect is significant, the Post Hoc Tests section has been opened. The relevant effect has been transferred to the right. The have used the Holm test; the range of tests and their potential usage will be discussed at the end of the chapter.

Post Hoc Comparisons - Phase

Comparison		Mean Difference	SE	df	t	P~holm~
Phase	Phase					
Pre - During		1.400	0.407	9.00	3.441	0.015
- Post		0.100	0.420	9.00	0.238	0.817
During - Post		−1.300	0.213	9.00	−6.091	< .001

While there is no relationship between the pre- and post- phases, there would appear to be significant differences between the campaign period and each of these phases. It's time for a chart.

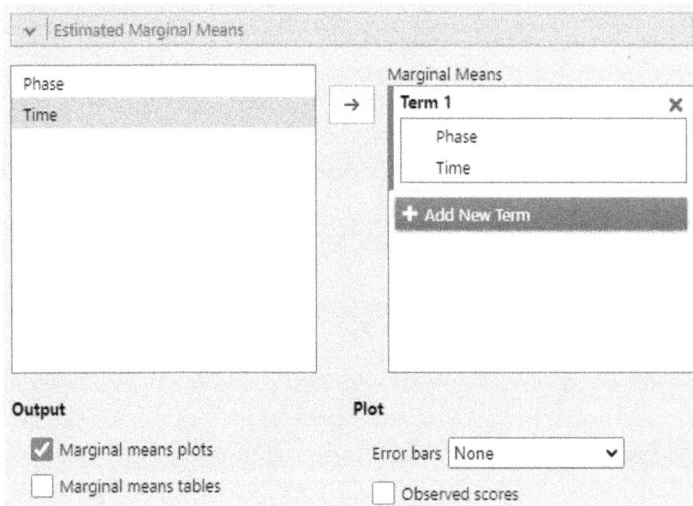

Although this may be useful for more detailed study, the Error bars option has been turned off for clarity.

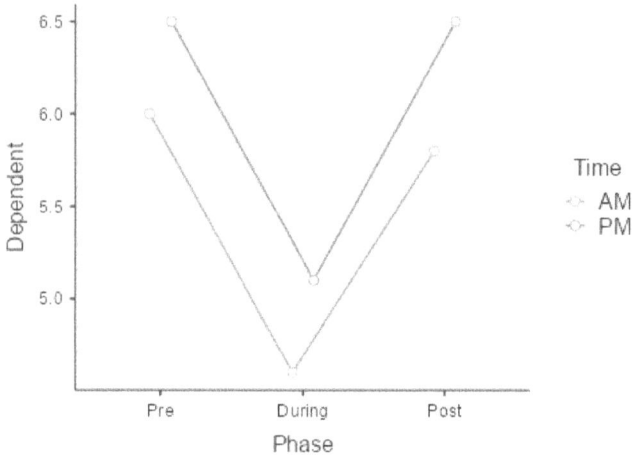

Three points arise from the chart. Firstly, we see that incidents are clearly lower during the campaign, but not only were they higher before, but they also rose to similar proportions afterwards. So the intervention would appear to have had an effect, but the stimulus did not have lasting effects. Secondly, the fact that both time lines (AM, PM) follow the same pattern indicates that the Phase factor is a 'global' effect. Thirdly, the lines do not intersect, nor even tending towards it; there is no sign of interaction between the factors.

While we are on the subject of interactions, it is worth knowing that not only is it possible to get a significant interaction, but there are also times when only the interaction is seen to be significant. In one (real) educational survey, the main effects on native students' performance outcomes were supposed to be the levels of immigration at each school and the average parental education level of the immigrant students. Neither main effect seemed to influence performance significantly. On the other hand, there was an interaction between the two effects. Apparently, the parents of immigrant students were quite often graduates.

Generally, it is worth saying that all such plots should be examined with care. You may also want to select Marginal means tables from the Output sub-section for reporting and for in-depth analysis.

Repeated Measures Three-Way ANOVA

Open the ThreeWayRepeatANOVA.csv file. Click the ANOVA tab and select Repeated Measures ANOVA. We are running a new program later in the year to deal with the same problem; maybe the lesson will sink in, or it's a different season, or maybe a new technique such as a video with role models may do the trick. The main point here is about data input.

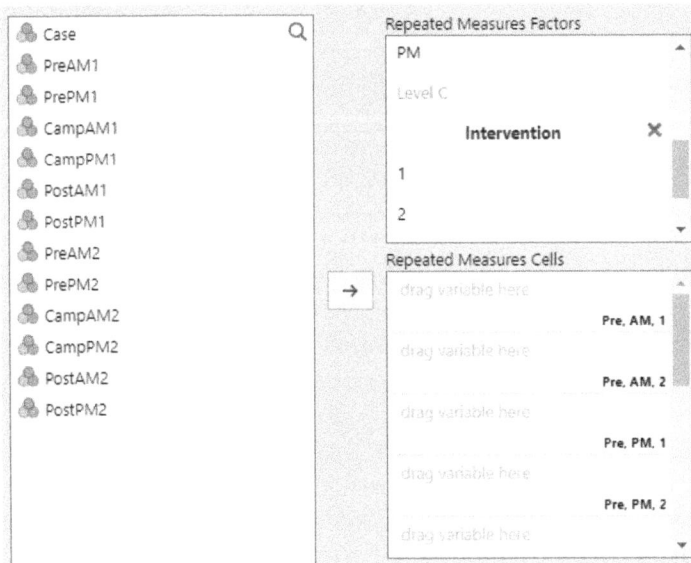

In terms of the source file, as you can see on the left-hand side, I have doubled the number of variables in the file, renaming the original ones with a '1' suffix, using the same names but with a '2' for the new set, representing the second intervention. Unlike the simpler two factor ANOVA, the variables do not match the right-hand permutations from top to bottom; *each variable needs to be put in carefully, one by one, to ensure correct matching.*

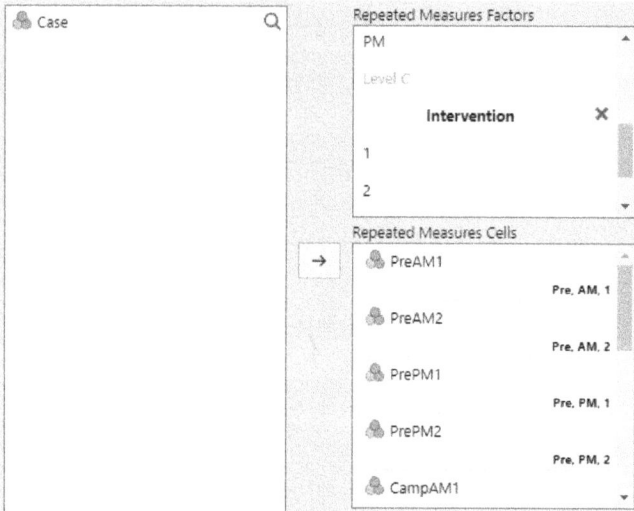

Here you can see some of the matches.

For Repeated Measures ANOVA, we can ignore the Between Subject Factors box, which is for the mixed design factorial ANOVA, to be discussed later in the chapter, where we have a hybrid of repeated measures (same subjects) and also between-subjects design (using groupings). Also ignore the Covariates box, which is for variables which we think are influential, but are not of interest to the study.

The output shows significant results for Phase again, but also for an interaction, Phase * Intervention, the latter being our new factor. You may want to check effect sizes. Charts from Estimated Marginal Means are called for:

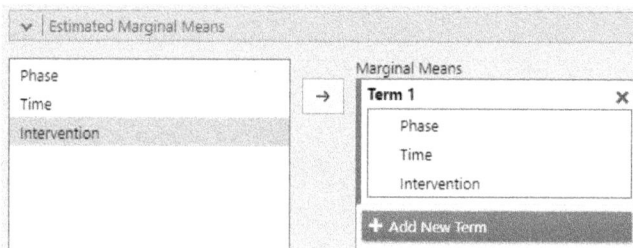

In order to examine the same effect as previously 'before and after', I have decided to have the same chart specifications as before, but viewing 'Intervention' as separate plots ('1' and '2').

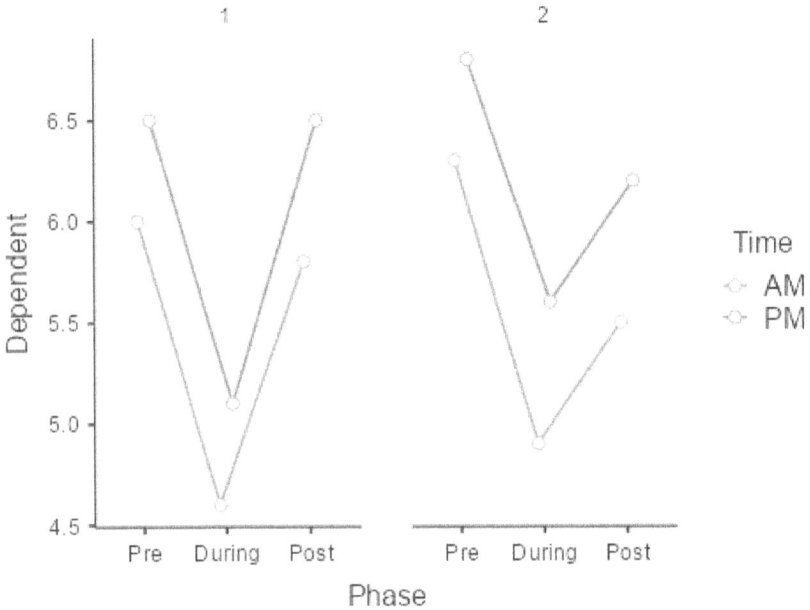

The charts suggest rather less of a bounce-back after the second intervention. On the other hand, there appear to be more incidents during the campaign than previously, and the post-campaign reduction is not as drastic as may have been hoped.

Let's see a chart for the interaction:

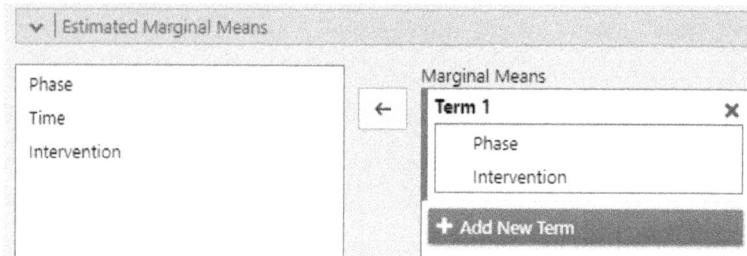

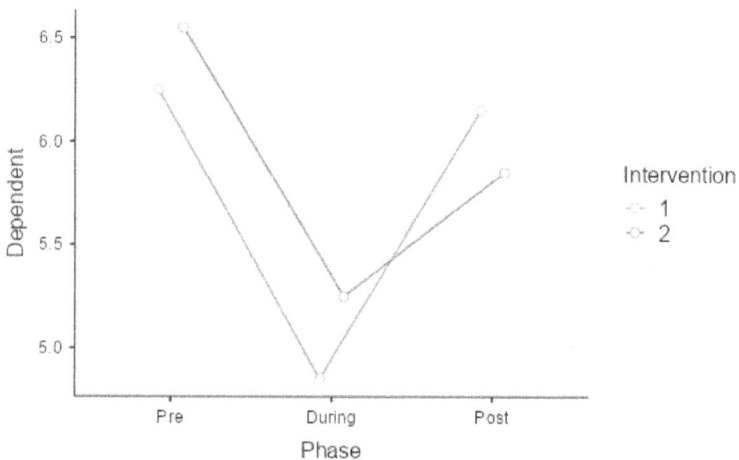

Note the intersection of the lines, indicating an interaction of the two effects. If we had a movement towards intersection rather than actual crossing, this would suggest a trend towards interaction. If you add Time to the interaction, the same effect will be apparent regardless of the different times.

Between-Subjects ANOVA

Open the BetweenANOVA.csv file.

Case	Coaching	Sport	Gender	Rating
18	A	Football	Female	7
19	A	Football	Female	6
20	A	Football	Female	8
21	B	Basketball	Male	3
22	B	Basketball	Male	4
23	B	Basketball	Male	6
24	B	Basketball	Male	1

We are interested in the effects of contrasting coaching styles on the motivation of sports professionals. Also of interest are gender and the difference between sports involving different levels of physical contact, say between basketball and gridiron (American) football.

Press the ANOVA tab, selecting ANOVA from the drop-down menu.

Here we have a three-way ANOVA. It is customary to check for similar variance across the groups, so I have chosen Homogeneity test (the Levene test).

ANOVA - Rating

	Sum of Squares	df	Mean Square	F	p
Coaching	47.433	2	23.717	6.8578	0.002
Sport	25.350	1	25.350	7.3301	0.009
Gender	6.017	1	6.017	1.7398	0.193
Coaching * Sport	4.300	2	2.150	0.6217	0.541
Coaching * Gender	0.233	2	0.117	0.0337	0.967
Sport * Gender	0.817	1	0.817	0.2361	0.629
Coaching * Sport * Gender	0.433	2	0.217	0.0627	0.939
Residuals	166.000	48	3.458		

Homogeneity of Variances Test (Levene's)

F	df1	df2	p
0.792	11	48	0.647

Levene's test is non-significant, so we don't have to worry about that. (If you chose the 'Q-Q', quantile-quantile plot, you would see that the points generally cling close to the diagonal line, indicating a normal distribution.) Here, only the Coaching and Sport variables are deemed significant.

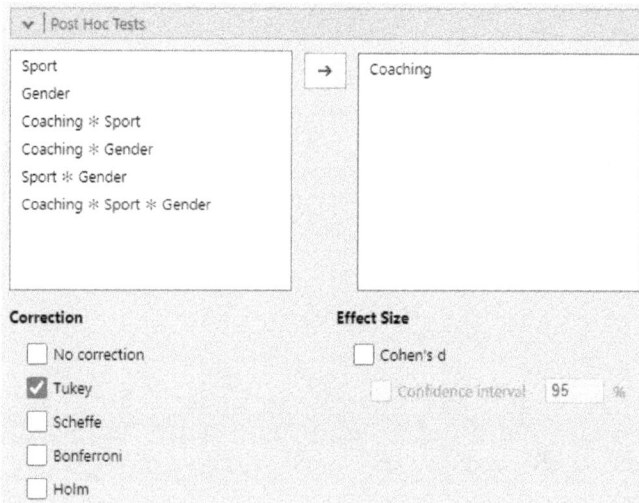

We only use a comparison test with Coaching as it has three conditions, so pairings make sense. I am interested in the effect of Coaching on motivation and am particularly interested in how they are considered by people from different sports. The Tukey test is the default here.

Post Hoc Comparisons - Coaching

Comparison						
Coaching	Coaching	Mean Difference	SE	df	t	P_{tukey}
A	- B	1.550	0.588	48.0	2.636	0.030
	- Control	2.100	0.588	48.0	3.571	0.002
B	- Control	0.550	0.588	48.0	0.935	0.621

Coaching method A differs significantly from the other two conditions.

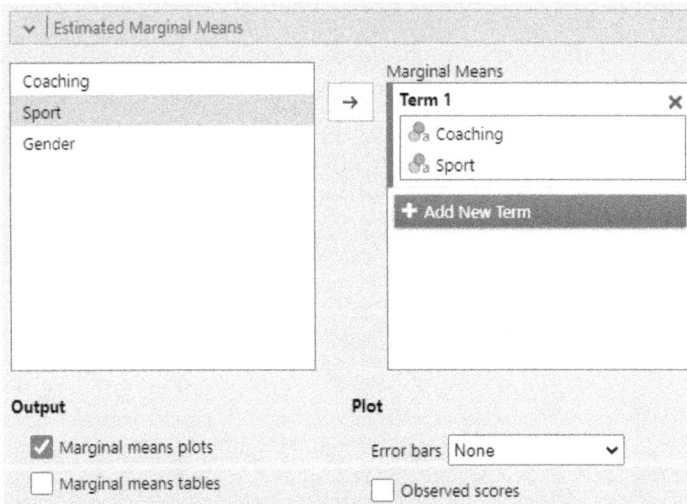

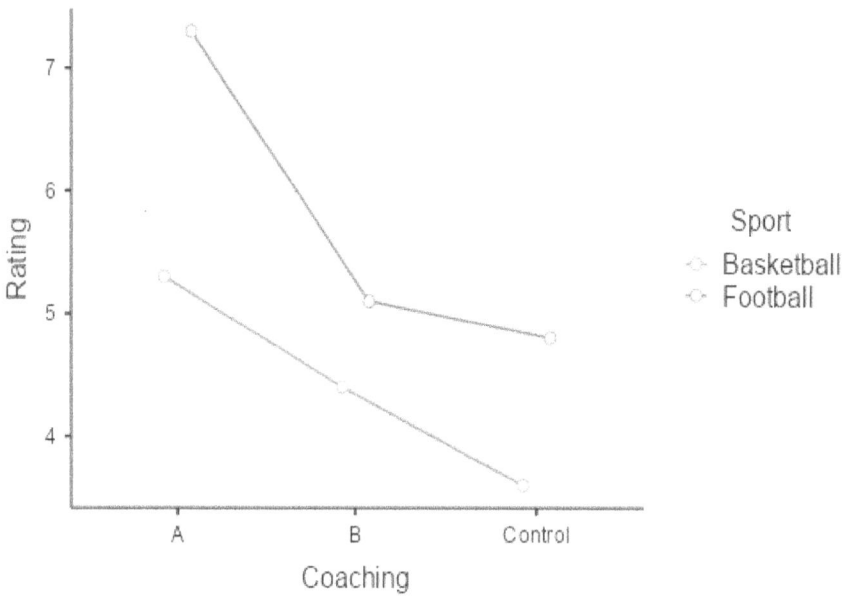

The chart shows quite a lot. Firstly, football players seem much more likely to be enthusiastic than basketball players, and this seems independent of the coaching methods. Looking at the leftmost node of the Basketball ratings, we can also see that Coaching method A elicits superior ratings regardless of sporting preference, as it rates higher than all B and Control nodes; this is supported by the comparison tests, with Coaching method A significantly different from B and from the Control result.

Mixed ANOVA

Open the MixedANOVA.csv file.

Case	School	Mathema...	English	Science
1	state	80	82	78
2	state	65	67	64
3	state	50	58	45
4	state	68	69	70
5	state	63	66	63
6	state	57	56	58
7	private	84	83	84
8	private	70	75	71
9	private	70	76	72
10	private	57	62	58
11	private	46	60	42
12	private	55	64	51

There is a grouping according to type of school, but all schools take examinations in the same subjects ('repeated measures'): Mathematics, English and Science.

Click the ANOVA tab and select Repeated Measures ANOVA from the drop-down menu.

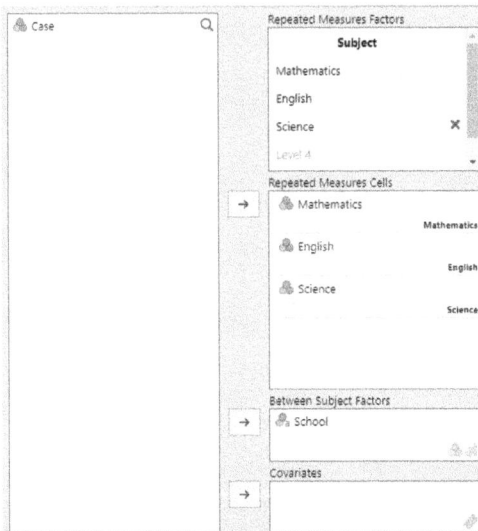

Here we have three subjects entered as Repeated Measures and one grouping in Between Subject Factors. This is a two-way mixed ANOVA, as there is one Repeated Measures (RM) factor, the academic subject and one Between Subject factor, the type of school. If you wanted to extend the design, there are two ways of creating a three-way ANOVA: you could put another grouping factor in the Between Subject Factors box, gender perhaps, or you could create another RM factor, maybe entering the results according to both final year and test results from a previous year. It is of course possible to have more than three factors, but the more of them you have, the harder it is to interpret the results.

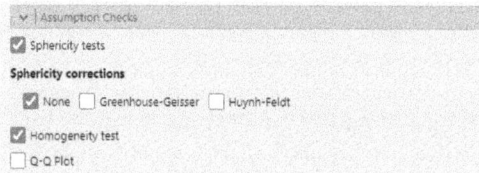

Because we have a between-subjects factor, we need the homogeneity test, checking that there are similar variances across the conditions. We also want the sphericity test; the sphericity assumption is rather like homogeneity of variance for repeated measures.

Tests of Sphericity

	Mauchly's W	p	Greenhouse-Geisser ε	Huynh-Feldt ε
Subject	0.161	< .001	0.544	0.559

Homogeneity of Variances Test (Levene's)

	F	df1	df2	p
Mathematics	1.075	1	10	0.324
English	0.355	1	10	0.564
Science	1.486	1	10	0.251

The sphericity test, Mauchly's test, is significant. We need to take a different reading of the results, providing a slightly more conservative p value to avoid Type 1 errors. Sphericity corrections are available in the

Assumption Checks section. Some researchers use the Greenhouse-Geisser correction on all such occasions, others use the Huynh-Feldt.

My advice is to return to the readings shown to the right of the Mauchly result; if the Epsilon statistic for Huynh-Feldt is greater than .75, use Huynh-Feldt; otherwise, use Greenhouse-Geisser (Girden 1992). So in this instance, we use the latter.

Sphericity corrections

☐ None ☑ Greenhouse-Geisser ☐ Huynh-Feldt

Within Subjects Effects

	Sphericity Correction	Sum of Squares	df	Mean Square	F	p
Subject	Greenhouse-Geisser	187.1	1.09	172.0	8.99	0.011
Subject * School	Greenhouse-Geisser	28.2	1.09	25.9	1.35	0.274
Residual	Greenhouse-Geisser	208.1	10.87	19.1		

Between Subjects Effects

	Sum of Squares	df	Mean Square	F	p
School	12.2	1	12.2	0.0313	0.863
Residual	3915.4	10	391.5		

The reading, as corrected, is shown, with an adjusted p value of 0.011 for the Subject main effect. Neither the school main effect nor the interaction is deemed to be significant.

As the Subject effect is of interest, we can go to the Post Hoc Test section to look at the pairings.

Post Hoc Comparisons - Subject

Comparison						
Subject	Subject	Mean Difference	SE	df	t	$P_{bonferroni}$
Mathematics -	English	-4.417	1.186	10.0	-3.72	0.012
-	Science	0.750	0.743	10.0	1.01	1.000
English -	Science	5.167	1.801	10.0	2.87	0.050

I have chosen to view the Bonferroni correction, rather than the Holm, for a reason. Given that our data is in breach of an assumption for repeated measures tests, it may be wiser to choose a more conservative test. It is to the subject of multiple comparisons that we now turn.

Multiple comparisons

Also known as post hoc tests, corrections or adjustments, these tests are designed to prevent Type 1 errors, the assumption of a significant result when none exists. The concern is that if you have several pairings, there is the chance that one or more results will only appear to be significant through a fluke. (Do note that the tests cited here are parametric tests; Chapter 5 includes comparisons to follow the non-parametric Friedman and Kruskal-Wallis tests.)

The multiple comparison tests may be used in a planned or unplanned way. With planned comparisons, you start with a well-considered analysis of variance, including the expectation that particular pairings will require subsequent analysis. With post hoc (after the event) comparisons, you follow up a large but significant ANOVA result with wall-to-wall coverage of all possible pairings. I have provided two rather extreme models, as everyday practice usually falls between these, but the principle is worth considering for the purposes of this discussion.

Before choosing between comparison tests, liberal and conservative, I would like to introduce the idea that we should use none. The position that the ANOVA is an 'omnibus' test from which only the overall result could be accepted was defended by the great statistician R. A. Fisher (1935). Criticism of comparison tests persisted for quite some time (Nelder 1971; Plackett 1971; Preece 1982).

Prior to the days of widespread computing, we did all the tests by hand, which was time-consuming but kind of fun (depending on one's view of life). Textbooks often didn't include multiple comparisons (for example, Greene and D'Oliveira 1982). Adding hugely to the amount of time expended on calculations, it is not too surprising that multiple test usage was not widespread before computer usage (Parker 1979), but well into the day of the personal computer, there was still the view that "multiple comparison methods have no place at all in the interpretation of data" (Nelder 1999).

Perhaps the widespread incorporation of comparison tests into computerized statistical packages has led to a general acceptance of their use, but here is one group of researchers fulminating against the 'wickedness' of the use of a liberal test: "the true interpretation of the data was submerged in the swamp of significance statements" (Mead *et al* 2012). The criticism of unreasoned usage still stands.

There is now little opposition to the use of planned tests. Where there are just a few anticipated pairings, it is generally considered reasonable not even to use the correction tests to be discussed in this chapter. Jamovi offers a 'No correction' option for ANOVA, or *t* tests could be used. If you are using the Friedman test, you can use Wilcoxon tests of each pairing; after a Kruskal-Wallis test, use Mann-Whitney tests.

Unplanned tests (*post hoc*) are still very controversial (Games 1971; Sato 1996). Pearce (1993) considers them overused, because of automated computerized usage (not in Jamovi) and a lack of desire to specify contrasts. While unplanned tests are said to be less powerful than planned tests (Day and Quinn 1989), I would suggest an intermediate view: unplanned tests could be used to generate new hypotheses, even though they should still be viewed with considerable skepticism. A more radical view is that of Hess and Olejnik (1997), that ANOVA should be abandoned in favor of focused hypothesis testing.

On the assumption that we wish to carry on using ANOVA and multiple comparison tests, we have another little controversy to consider. In this book, I have only followed up significant ANOVA results with comparison tests. Many consider follow-up from non-significant ANOVA tests to be a form of 'dredging' (for example, Wyseure 2003). This view is challenged as a misconception by other researchers (for example Huck 2008), who believe that comparison tests may be used regardless of whether or not the overall effect was significant.

Having considered these matters, let us consider the tests themselves. Some of the time, the different tests will give similar results. At other times, they differ. The following descriptions are only general; because the tests use different techniques, their apparent 'liberalism' or 'conservatism' should not be viewed as forever accurate.

In general, the **Holm** is considered the most liberal of the tests provided (also known as 'more powerful'); it is most likely to provide a lower p value, and thus more likely to pick up 'significant' results. This also means that it is slightly more likely to commit a Type 1 error, wrongly attributing significance to an effect. The Holm can be used for both Repeated Measures and between-subjects ANOVA.

The **Scheffe** is generally the most conservative of the tests. It will avoid Type 1 errors easily, but can occasionally be prone to Type 2 errors, wrongly considering results to be non-significant. Even if you have chosen this test, it is probably also worth checking the results of a Bonferroni test at the same time; every now and again, the Scheffe can be completely adrift from other test results. Results quite often differ, but they should not differ wildly. The Scheffe should only be used for between-subjects ANOVA.

The **Bonferroni** is a traditional conservative test. While not as tough-minded as the Scheffe, it is considered to be a little too strict for many occasions (Rice 1989), although is still used in several textbooks (for example, Kinnear and Gray 2004). The Bonferroni can be used for both Repeated Measures and between-subjects ANOVA.

Tukey is a reasonable intermediate test, not as strict as the Bonferroni, but not as liberal as the Holm. Traditionally used and the default for this book, the Tukey is the most widely used test (Tsoumakas *et al* 2005). The Tukey should only be used for between-subjects ANOVA.

Dallal (2012) suggests the use of non-adjusted tests such as t tests for planned tests. Scheffe could be used where you want to use all of the comparisons you can think of, but Dallal is scathing about possibly missing an effect for the sake of comprehensiveness. Tukey is the comparison test he most often uses.

Hilton and Armstrong (2006) stress the importance of the purpose of the investigation in deciding upon test usage. "If the purpose is to decide which of a group of treatments is likely to have an effect, then it is better to use a more liberal test.. in this scenario it is better not to miss an effect. By contrast, if the objective is to be as certain as possible that a particular treatment has an effect then a more conservative test.. would be appropriate." Hilton and Armstrong consider Tukey's test to fall between the extremes.

They also believe (as does Dallal 2012) that "none of these methods is an effective substitute for an experiment designed specifically to make planned comparisons between the treatment means". You have probably heard this before, but it is always worth keeping in mind, that good research design makes analysis a lot easier and more efficient.

Thinking point

Are you using a test, or any other form of technology, just because you can? Apart from advantages and disadvantages, irrelevances can obscure more important issues.

Chapter 11 – ANCOVA considered

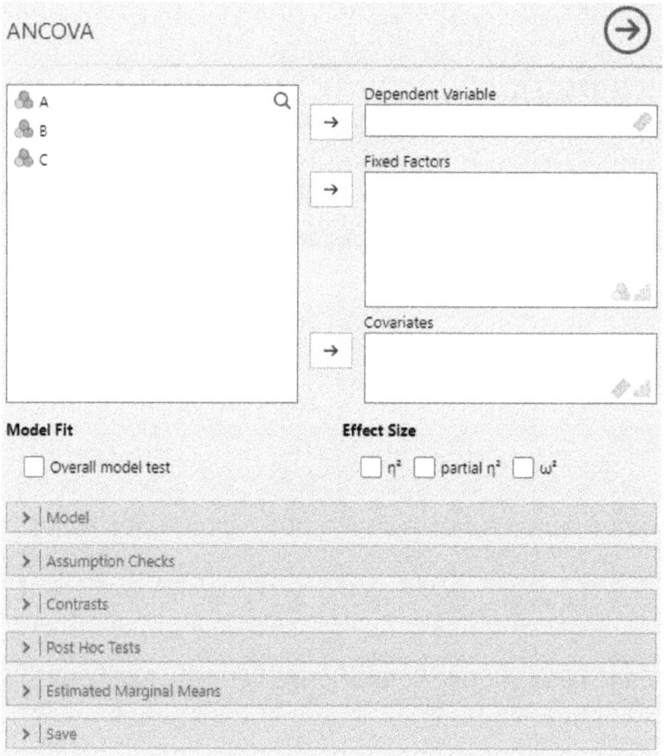

What ANCOVA does and what it is supposed to do

As is suggested by the Dependent Variable and Fixed Factors boxes, ANCOVA is quite similar in function to ANOVA (note that it is for a between-subjects design, not repeated measures, as error terms need to be independent). The Covariates box shows the difference: it is possible to enter other (continuous) variables in order to control for those variables. When you read a report about a study controlling for this and that variable, it means that the researcher wants to allow for the effects of unwanted variables on the main focus of study. Variables are likely to overlap.

ANCOVA – analysis of covariance – is in fact a combination of ANOVA and regression, with covariance being the extent to which variables change together.

In practice, you would get a read-out of values similar to those from ANOVA, but with values for the factors and the covariates. The most important point is that the values for the factors will have changed.

The purpose is to get more accurate estimates of a factor's influence on the variance. ANCOVA has also been used to try to remove the effects of fixed groups from a study (for example, administrative, managerial and manual staff) in order to study the overall effect. In both cases, ANCOVA is designed to remove statistical noise. But before you rush off and use this, I would first of all like to introduce some methodological issues which may make you think again.

Assumptions for data

There are the usual assumptions for using parametric data, those of continuous data, normal distribution and, for different numbers in different conditions, homogeneity of variance.

One additional assumption is the need for the covariate to have a linear relationship with the dependent variable; this can be measured using a scatter plot.

Another assumption, arguably the most important of the additional assumptions, is that of homogeneity of regression: the dependent variable and the covariate must not be over-correlated (yes, this test wants to have its cake and eat it; Goldilocks' approach to porridge comes to mind). Regression lines for the covariate across the different groups need to be parallel, neither crossing each other nor getting too close to each other.

The covariate should be unrelated to the dependent variable; this should be checked during the design stage, not as a result of the test itself.

If there is more than one covariate, these should not be over-correlated.

ANCOVA problems

As is suggested by the above, suitable data sets are quite narrowly defined. While proponents of most parametric tests cite their robustness, there is evidence to suggest that ANCOVA is "a delicate instrument" (Huck 2012). Serious critiques of ANCOVA describe problems relating to data reliability and the smoothing out of differences between mixed groups (Campbell 1989; Buser 1995; Miller and Chapman 2001). Huck (2012) blames the users! He feels that they often consider complexity to be a virtue in itself. His more measured description is thus:

> "To provide meaningful results, ANCOVA must be used very carefully – with attention paid to important assumptions, with focus directed at the appropriate set of sample means, and with concern over the correct way to draw inferences from ANCOVA's F-tests. Because of its complexity, ANCOVA affords its users more opportunities to make mistakes than does ANOVA." (Huck 2012)

Note that drawing inference, how you interpret the data, is also problematic. Reputable researchers have created serious flaws in their studies by using misleading results from ANCOVA (Campbell 1989 and Buser 1995 cite instances from educational research).

This book is aimed primarily at beginners and intermediate users. I would even suggest that advanced statistical users think carefully before using ANCOVA.

Alternatives to ANCOVA

One possibility is just to use *t* tests and ANOVA, recording the possibility of mediating factors. This may not be ideal, but to my mind is preferable to getting totally wrong results.

Another possibility is stratification. The data can be broken up according to groupings of different levels of the covariate (for example, status bandings), using the categories as 'fixed factors'.

Another possibility is to re-examine the model using multiple regression.

Thinking point

Is complexity a virtue? Is 'powerful' always a positive or even a meaningful adjective? Will the writer of these comments ever emerge from their cave?

Chapter 12 – MANOVA

MANOVA, multivariate analysis of variance, works like ANOVA, but it allows you to examine effects in the light of more than one dependent variable. The minimum number of independent variables is 1, essentially a t test with multiple dependent variables. The minimum number of dependent variables (sorry, obvious) is 2.

There are two reasons for running a MANOVA rather than individual ANOVA tests. Firstly, it removes redundant information. If two different measures both measure the same concept, then it could well be that simply using one dependent variable may be providing a rather rough and ready explanation. Indeed, it is possible that by doing so, you may miss an effect.

Secondly, we may learn about how different groups differ by examining their **linear composite**, an individual combination of dependent variables. Each group may differ in how its combination is constituted, offering us more of an understanding of the independent variable (Weinfurt 1995).

Also, by saving yourself from running a range of ANOVA tests, you are cutting down on the number of tests run. This decreases your chances of committing a Type 1 error (Weinfurt 1995), believing in a result which is in fact a fluke.

An example could be a study of the differences between university drop-outs and those who succeed (chin up). It is probably not caused by a single factor, and the relevant outcome measures may be interlinked (real research, by the way, indicates that a major issue is the specific academic subject). Another example is that of suitability of students for particular courses, based on the performance of previous students; it is often found that specific groups of aptitudes relate to success on a particular course of study, not one particular aptitude. So looking at more than one outcome measure may well tell us more about what is going on.

Assumptions for data

MANOVA is widely considered to be quite robust. As well as the usual assumptions for parametric data that we have already considered, the assumptions below should be adhered to as much as possible. Some tailoring of the data may be necessary to make it a more powerful and reliable instrument; one thing that will emerge from a study of the assumptions is that having equal numbers of cases in each sample, or near equal, is helpful.

Arising from the point made earlier about linked outcomes is the need for **correlated dependent variables**. There is some controversy about just how correlated variables should be. On the whole, there is consensus that high negative correlations are particularly good, as are moderate positive correlations. Highly correlated dependent variables are considered wasteful by some; you could consider replacing them with composite measures derived from variables with very similar underlying meanings (Tabachnick and Fidell 2013).

A really important assumption is that observations are independent. This is a matter of design. You need to be sure that people (assuming your subjects are human) did not overhear each other, see the results of others or otherwise contaminate the evidence.

Another assumption is multivariate normality. This is checked for in Jamovi by the multivariate Shapiro-Wilk test – reviews indicate that this is at least as good if not better than more traditionally cited tests (Alva and Estrada 2009; Yap and Sim 2011) – and also the Q-Q plot of multivariate normality.

Yet another assumption is homogeneity of variance-covariance matrices, a counterpart to the ANOVA homogeneity of variance. If you have equal sample sizes, then MANOVA is reasonably robust to this unpronounceable assumption and to normality (Kinnear and Gray 2008). The test for this, Box's M test, is notoriously sensitive. To give yourself some leeway on this test, you can set yourself a critical value of $p < .001$ (Tabachnick and Fidell 2013). If you still have problems, you could use a more conservative MANOVA statistic (to be mentioned when we run the tests).

The test

Open the Manova.csv file and press the ANOVA tab, selecting MANCOVA from the drop-down menu.

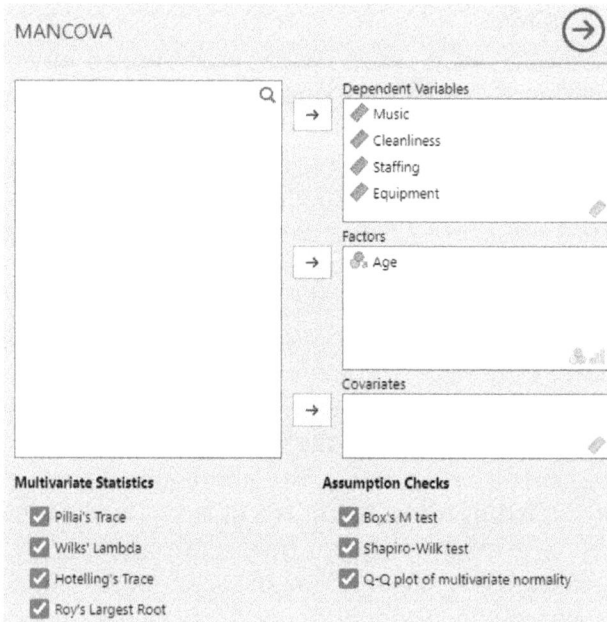

The four dependent variables are measures representing the views of the users of the gym at a fitness center: the music, cleanliness, levels of staffing, and the quality of the equipment. Recent feedback results have been inconclusive, but the researcher wonders if the age of users may be an influential factor. The independent variable comprises five relatively arbitrary age bandings: 61+, 51-60, 41-50, 31-40, 18-30. This reminds me: it always saves confusion when creating a questionnaire if you ensure that each banding is exclusive; the number of times I've seen things like 40-50 and 50-60, giving 50-year-olds two richly deserved bites of the cake. *

Assumption Checks

Box's Homogeneity of Covariance Matrices Test

χ^2	df	p
45.6	40	0.250

Shapiro-Wilk Multivariate Normality Test

W	p
0.984	0.080

Box's M test for homogeneity of variance/covariance matrices. As this test is super-sensitive, we only need to be >= 0.001 (Tabachnick and Fidell 2013). We have a large p value of 0.25, so no worries there.

The normality test has a p value of 0.08. It's greater than .05 so we don't have evidence with which to reject the null hypothesis. A significant result would have indicated likely abnormality of the distribution within the data set.

---

*The dataset is a transformed version of the Egyptian skulls data from Thomson and Randall-Maciver (1905).

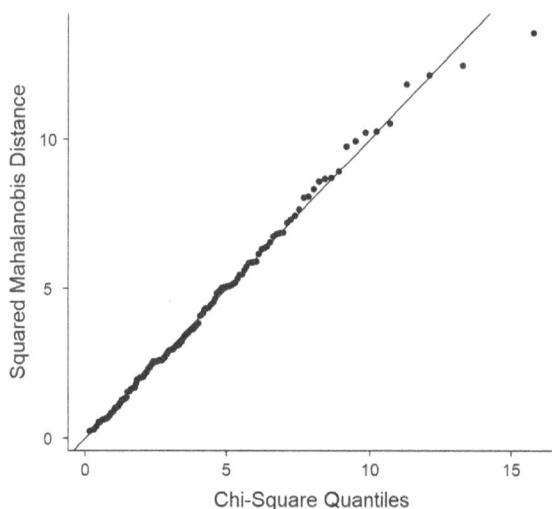

The 'quantile-quantile' plot provides a graphical assessment of multi-variate normality. The points form a straight line, more or less.

Multivariate Tests

		value	F	df1	df2	p
Age	Pillai's Trace	0.354	3.52	16	580	< .001
	Wilks' Lambda	0.664	3.90	16	434	< .001
	Hotelling's Trace	0.481	4.22	16	562	< .001
	Roy's Largest Root	0.422	15.3	4	145	< .001

On this occasion, I suggest looking at the default display of all four statistics to get a feel for what they look like. Generally, these give similar results; in this particular case, the p values all reach the same critical value, $p < .001$. Clearly, age differences have an effect on the dependent variables.

Where there are only two levels to an effect, the statistics should be identical. Otherwise, they are usually similar. However, the Pillai is considered very conservative, so is probably the best to use if you have some problems with homogeneity (be reasonable – you can't stick in any old data; the MANOVA will give you a result with its poker face and you won't know if it's just bluffing). Wilks' lambda is the most widely used, so is probably a reasonable default. Roy's has the reputation of being the most liberal test, although Hotelling's test is also supposed to be quite liberal. I would suggest Roy's if you have data derived from an experiment which also comfortably passes the multivariate normality, Hotelling's for non-experimental data which nevertheless passes the assumptions tests with ease. On this occasion, with some doubts about normality, I would suggest Wilks' lambda. However, the problem is that none of these tests conform to type in all situations, essentially because they calculate things differently and were not necessarily designed for the purposes of taking a stand against each other in terms of their social attitudes!

Anyway, if you were to report Wilks' lambda, this would include the statistic itself, 0.664, the F value 3.9 and the p value, here so small that it is recorded as < .001. There is a significant effect involving a combination of dependent variables.

Univariate Tests

	Dependent Variable	Sum of Squares	df	Mean Square	F	p
Age	Music	0.00535	4	0.00134	6.01	< .001
	Cleanliness	0.00251	4	6.27e−4	2.46	0.048
	Staffing	0.01626	4	0.00406	8.28	< .001
	Equipment	0.00443	4	0.00111	1.50	0.207
Residuals	Music	0.03228	145	2.23e−4		
	Cleanliness	0.03690	145	2.55e−4		
	Staffing	0.07113	145	4.91e−4		
	Equipment	0.10748	145	7.41e−4		

This gives us an ANOVA for each of the separate dependent variables. Music and Staffing are obviously valuable outcome variables when we consider age differences, with a rather more tenuous result for Cleanliness and no particular significance for Equipment.

There is considerable debate about whether or not you should adjust the ANOVA results for Type 1 errors. A review of the evidence in Weinfurt (1995) suggests that while MANOVA controls for Type 1 errors in the situation of null significance, this may not be so in the case of a significant result.

If you do adjust your results, then the old-fashioned way of doing this is to use a Bonferroni adjustment. This involves counting the number of hypotheses: if you have three variables, that means three pairings; with four variables, this expands to six pairings.

Let us say that we have already decided to ignore the Equipment variable but wish to consider the rest (I'm feeling generous). So, we have three variables, with just three pairings.

The simplest way of carrying this out (if not the most correct way) is to multiply the p values by the number of pairings and then to compare them with your critical value (we assume $p < .05$). So we have Cleanliness = 0.048 x 3 = 0.144, which is well over the 0.05 critical value. Bonferroni is these days considered a bit harsh (although I'm sure he was a nice bloke). Let us instead try something a little more sophisticated, Holm's sequential Bonferroni (Holm 1979).

A slightly more time-consuming but more liberal (or 'more powerful') adjustment, the Holm-Bonferroni still starts with working out the number of pairings. You start with the most significant result and multiply by the pairing number: Staffing .001 x 3 = .003. Then take the next highest result and multiply by one less: Music .001 x 2 = .002. And so on: Cleanliness .048 x 1 = .048; still < .05.

A follow-up routine could be to break up the data set into subgroups according to age group (try Descriptives) and then see if there are differences between the categories. You can take it from me that the 61+ age group sees things differently from the 18-30 group, and that's not just about the night life!

To examine group separation within a dependent variable without breaking up the groupings, I would suggest using logistic regression. This is covered in Chapter 14.

A note on MANCOVA

The Jamovi team have chosen to place MANOVA within MANCOVA because the additional box (Covariates) allows MANCOVA to be used. If you are interested – and are willing to be warned – read the previous chapter. Obviously, the same additional assumptions and pitfalls apply to MANCOVA as to ANCOVA.

Part 4

Relationships, broad and narrow

Chapter 13 – PCA and factor analysis

Introduction

Factor analysis (FA) is a set of approaches and techniques for data reduction. That this definition can be challenged is left until later in the chapter, but in general I think it reasonable to say that we conduct some form of FA to make sense of multiple variables by grouping them together and reducing them to a smaller set of entities, thus reducing redundancies of correlated variables.

There are three main approaches within FA, each conventionally using a specific set of techniques. Exploratory factor analysis (EFA) attempts to find out what the variables have in common, producing smaller, more meaningful conglomerations called factors. Confirmatory factor analysis (CFA), which follows a similar set of statistical assumptions, tests the factorized models that have already been developed. Principal components analysis (PCA) condenses multiple variables into fewer conglomerations, this time called components, but meaningfulness is not theoretically built in. To illustrate the difference, let us consider three roughly drawn research instances.

Let us say that you have a large number of items on a personality test that you are developing, perhaps organized in Likert scale format. You may seek a way of finding out which questions could be eliminated or perhaps consolidated into core questions. This sort of straightforward condensation is almost invariably conducted by principal components analysis (PCA). While you could do this by 'eyeballing' a correlation matrix to seek similarities, it is a lot easier to use PCA. Once you have a good idea of the components, you may then find the matrix more meaningful.

If you are interested in how many key ideas underlie different variables pertaining to a social issue, what they mean and how the ideas interrelate, then exploratory factor analysis (EFA) has traditionally provided the toolkit of choice.

If you have an established model and you want a verification procedure to test it using different data, or to compare it with other models, then confirmatory factor analysis (CFA) is a traditional approach. This is not covered in this book and we will restrict our discussion to PCA and EFA within a general term called 'factor analysis'.

In principle, PCA has a different philosophy from EFA (and CFA). PCA is empirical: constrained by the data, it has no underlying model and calculates a number of components on an automatic basis. EFA is studiedly theoretical; in its purest form, the researcher has a theory, considers a model (or models) and examines their fit with the data. Let's go into a little more detail.

PCA and EFA, compared and contrasted

Both approaches create linear combinations from the variables in a dataset (Stevens 2009). These are not completely different in concept from the linear composites found in multiple regression.

Principal components analysis is the product of statisticians, and it shows. The concept was that of Karl Pearson in 1901, developed and named in the 1930s by another statistician, one Harold Hotelling (the sharp-eyed will find him in a walk-on role in the MANOVA chapter). Their approach is mathematical, essentially data-driven (bottom-up rather than top-down, to use a psychological concept). The PCA process tells us, or tries to tell us, how many common features can be detected. These are called **components** rather than factors. We do not get to specify the number of components, as there is no underlying theory to take into account (Tabachnick and Fidell 2007). This is an empirical approach: the machine works it out for us, or tries to, indicating the most likely number of components which can be condensed from the variables.

Factor analysis, in its purest sense, is a theory-driven approach, the invention in 1904 of another friend of ours, psychologist Charles Spearman. Its purpose is to seek underlying meanings. The **factors** are also called dimensions, latent variables or even constructs (psychologists at work). We start with a theory, or rather a model representing that theory, decide that a number of constructs should exist, and then we choose which number of factors to examine at the beginning of the process. We might in some studies try out a different number of factors as well, but that is because we are open to another interpretation, not just because we want to see what happens. Underlying causes are assumed, with factors 'causing' the variables (Tabachnick and Fidell 2007).

Both PCA and EFA as techniques decompose a correlation matrix into constituent factors. PCA at its most basic just transforms the observed correlation matrix into components using the total variance, creating a linear function called the eigenvector; the eigenvector's value, known as the eigenvalue, explains the total variance of the component. Factor analysis focuses on common variance between the variables (covariance), seeking 'communality'; this is a mathematical estimate (Stevens 2009). Its calculations are rather more complex than those of PCA; you will of course be pleased to know that a detailed explanation can be found in Bryant and Yarnold (1995).

At the other end of the calculations, each column representing a component or factor contains **loadings**. These statistics represent the strength of the relationship between the variables and the component/factor, but their initial state is rather nebulous.

After the transformation of the data, both sets of techniques can be subjected to **rotation**, with the transformed data being spun around on its axis. This does not alter the data but allows us a clearer view, seeking what is known as **simple structure**, a clearly definable set of components or factors.

I think a useful analogy for rotation is looking up at the stars. The constellations are only groups of stars from the point of view of an observer here on Earth, not an objective reflection of their locations in space; similarly, a view from another part of the galaxy would be just another viewpoint. Rotation gives us another way of viewing the factors or components.

Traditionally, EFA has been the home of rotation, but it is now quite common for principal components analyses to be accompanied by rotation, as is the case in Jamovi (based on implementations in the R programming language). Some would say that the combination of principal components analysis with rotation should no longer be called PCA but rotated PCA, using the name of the rotation technique; if you use the varimax rotation, for example, this would be 'varimax-rotated PCA'.

This final point demonstrates the considerable blurring between the two apparently different approaches. Another example is that some people conducting factor analysis (FA) are known to use PCA to decide on the number of objects before conducting factor analysis; yes, I know, what price theory when you do that? To my mind we can use either approach for 'factor analysis' if we choose to use the term as a general methodology rather than as a group of techniques. This will be discussed later, after we have had a little practical experience.

Assumptions for data

Traditional PCA is a non-parametric procedure, with fewer assumptions to be met than for factor analysis, but it requires a linear relationship between all observed variables, with either continuous data or well-calibrated ordinal data. *

---

*Item response theory is recommended for the calibration of Likert scales in pilot projects.

Traditional EFA has more assumptions for the data to meet: normality, linear relations between the variables, correlations between several variables, an absence of outliers, sampling adequacy (as measured by Bartlett's test and KMO) and a reliable sample size.

The Factor module uses implementations of Horn's parallel analysis (PA) as its default algorithm. PA is very robust and largely impervious to the type of data distribution (Dinno 2009). I would still recommend some care about the data you put in; remember 'GIGO', garbage in, garbage out.

On the other hand, sample size is still something of an issue. Traditionally, 100 was seen as the absolute minimum number of cases, although some authorities set this at 50, with up to 200 considered not very good, stretching into thousands before the pundits were happy. There was also a stipulation about the proportion of variables to cases. It was proposed that a minimum of 5 cases was needed per single variable (Gorsuch 1983), but with a far larger ratio of observations to variables if the results were critical. To throw the fat onto the fire, it was found that even combinations of these rules were unpredictable and researchers began to use tests such as the KMO (Kaiser-Meyer-Olkin), which is available in Jamovi. Even if it were not, there is evidence that the parallel analysis algorithm offered in Jamovi is also quite robust when it comes to sample sizes (Ladesma and Valera-Mora 2007).

In spite of this, sample size should not be ignored. As I am sure is obvious, larger samples are better for the purposes of generalization and replication, but size also has a considerable effect on the number of factors or components indicated. This latter point will be discussed further when we reach the choice of methods for calculating the likely number of components.

The effectiveness of parallel analysis

Horn's parallel analysis (PA) is based on Monte Carlo simulation methods, which calculate by using multiple random samples from the data.

Because PA is labor-intensive, it takes a relatively long time to work, only recently coming into its own with the advent of fast personal computers.

PA is generally considered to be a more accurate way of calculating the number of factors in a data set than traditional PCA and FA algorithms. This certainly seems to be the case with small samples, where the PA mode beats the Kaiser criterion and the scree plot as arbitrary ways of choosing the number of factors (Ladesma and Valera-Mora 2007). However, this seems not to be the case when observations run into the hundreds; fortunately, the scree test, also provided by Jamovi, becomes a more efficient arbiter with larger data sets. In the meantime, rather than throwing more information at you, I intend to dive in and explain details as they appear.

Principal components analysis in action

PCA can be used for a wide range of purposes. You may want to reduce multiple measures of fitness, individuals' results on a battery of personality and aptitude tests, or most typically, reduce the number of survey questions to a more manageable number. (Beware of asking too many questions: this reduces the number of people providing you with answers!)

I have provided a very small file (in factor analytic terms) called PCA.csv, which contains 12 variables and 50 cases. Press the Factor tab and select Principal Component Analysis from the drop-down menu. Shift the 12 variables to the right-hand Variables box:

Principal Component Analysis

Variables

- A
- B
- C
- D
- E
- F
- G
- H

Method

Rotation | Varimax

Number of Components

- ◉ Based on parallel analysis
- ◯ Based on eigenvalue

  Eigenvalues greater than | 1

- ◯ Fixed number

  1 | component(s)

Assumption Checks

- ☐ Bartlett's test of sphericity
- ☐ KMO measure of sampling adequacy

Factor Loadings

Hide loadings below | 0.3

- ☐ Sort loadings by size

Additional Output

- ☐ Component summary
- ☐ Component correlations
- ☐ Initial eigenvalues
- ☐ Scree plot

> | Save

The current configuration is the default. The Parallel Analysis option chooses the number of components automatically, here with an orthogonal rotation method called varimax (discussed later). On the screen to the right, you will see the results:

Principal Component Analysis

Component Loadings

| | Component | | |
	1	2	Uniqueness
A		−0.929	0.1375
B		−0.894	0.1707
C		0.588	0.6535
D	−0.390	0.393	0.6930
E		0.778	0.3488
F	−0.330	0.747	0.3329
G	0.732		0.4324
H	0.883		0.2016
I	0.899	−0.378	0.0485
J	0.774		0.3985
K	−0.932		0.0457
L	−0.925		0.1226

Note. 'varimax' rotation was used

Looking at the table, we see four columns. The column on the far left represents the variables. Any central columns represent the components discerned from the data; these show the loadings, the strength of the relationship between variables and the particular component. The right-hand column, Uniqueness, shows how far each variable's variance is unique to itself, and thus is not shared with other variables; high uniqueness means limited relevance to the components.

Of these statistics, the most reportable are the loadings, which would be shown within a matrix of the component headings and the variable names. By squaring a loading, by the way, you get the percentage of explained variation, the extent to which the variation of the variable is explained by the component (a bit like squaring a coefficient to get the effect size). So the higher the loading, the better.

In this example, PCA (or more specifically, varimax-rotated PCA using parallel analysis) indicates that there are two components. This is quite clear. If you see a component consisting of less than three variables, it is likely to be unstable; that's not a problem here. However, there is some lack of clarity. Don't worry about the negatives; dimensions can be defined by their opposites. Of more concern are the shared loadings, as is currently the case with variables D and I, and relative low loadings and high uniqueness, in particular variable D.

Turn on the assumption checks:

Assumption Checks

Bartlett's Test of Sphericity

χ^2	df	p
691	66	< .001

KMO Measure of Sampling Adequacy

	MSA
Overall	0.539
A	0.539
B	0.387
C	0.620
D	0.324
E	0.399
F	0.456
G	0.361
H	0.566
I	0.721
J	0.746
K	0.603
L	0.830

Bartlett's test for sphericity rejects the null hypothesis that all the correlations tested simultaneously are 'not statistically different from zero'. More prosaically, it needs to reject the null hypothesis in order for us to run a factor analysis. A 'significant' result with Bartlett's test does not guarantee a good analysis but is a form of safeguard against optimistic but meaningless hunts through data. The test is always significant with large data sets, but is useful with smaller data sets (Pedhazur and Schmelkin 1991). The result here is a very small p value; it seems acceptable to carry on with the analysis.

The MSA figures, measures of sampling adequacy, should be at least .5, which is not that good anyway, some would say .6, and the nearer to 1 the better. Here, we see that the overall statistic is not particularly good and that individual variables are well below these levels. What we need to do is to remove the lowest scoring variable and then reassess the situation, removing another variable as necessary.

We firstly need to remove variable D as it has the smallest MSA score (0.324); transfer it back to the left-hand box. You will see that this improves matters somewhat:

KMO Measure of Sampling Adequacy

	MSA
Overall	0.632
A	0.543
B	0.507
C	0.460
E	0.492
F	0.510
G	0.501
H	0.798
I	0.692
J	0.828
K	0.775
L	0.733

However, the MSA figures are still rather low. Remove the next smallest, variable C (0.460).

KMO Measure of Sampling Adequacy

	MSA
Overall	0.667
A	0.539
B	0.582
E	0.443
F	0.584
G	0.533
H	0.810
I	0.737
J	0.787
K	0.808
L	0.746

This is better, but variable E is still under .5; remove it.

KMO Measure of Sampling Adequacy

	MSA
Overall	0.745
A	0.672
B	0.594
F	0.635
G	0.780
H	0.854
I	0.745
J	0.755
K	0.753
L	0.814

The overall score is quite high and those for the individual variables are acceptable.

Component Loadings

	Component		
	1	2	Uniqueness
A		0.906	0.1792
B		0.930	0.1167
F	−0.347	−0.828	0.1945
G	0.728		0.4225
H	0.892		0.1646
I	0.904	0.336	0.0695
J	0.791		0.3701
K	−0.932		0.0684
L	−0.934		0.1005

Note. 'varimax' rotation was used

We now have a fairly clear pair of components. Variables G to L are loaded onto component 1; A, B and F are loaded onto component 2. We are not worried about negatives, as constructs can be built of opposites. We ignore the smaller loadings, although for the purposes of neatness, we can 'disappear' them from our matrix by raising the 'Hide loadings below' option to 0.4.

One of the reasons for using factor analysis, using the term broadly, is that this makes it much easier to see the wood from the trees than just looking at a swathe of correlations. However, now you've got some idea of what you are looking at, you may choose to go to a correlation matrix to examine the strength of relationships between specific pairs of variables – assuming that it helps you in some rational way. To do this, select the Regression tab at the top of the program and select the Correlation matrix from the drop-down menu. Shift the variables to the right as usual, but not variables D, C or E. Choose just one correlation method (Pearson, Spearman or Kendall). Personally, I would turn off the 'Report significance' toggle.

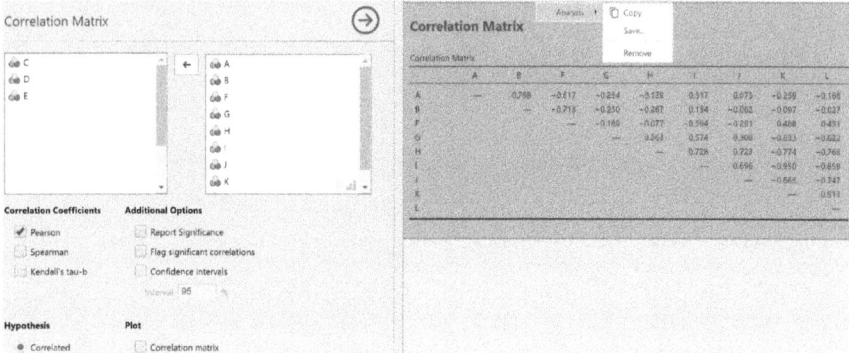

By right-clicking the correlation table, you can 'Copy' the table into the computer's memory. Then paste the results into a spreadsheet. To make it easier to see the stronger correlations, you may choose to use the spreadsheet's conditional formatting. A traditional cut-off for correlations in principal components analysis is above 0.3, but remember to account for negatives: you might specify Red < −0.3 and Green (or Blue!) > 0.3.

By this time, if you were narrowing down a battery of survey questions, you would have carried out the equivalent of a traditional principal components analysis. You will have been able to remove a question which was too similar to others to remain or created a new question standing in for others.

Traditional principal components analysis

Two older ways of working out the number of factors are the scree plot and the Guttman-Kaiser criterion (also known as the Kaiser criterion or Kaiser rule). If you go to the Additional Output options (assuming that you have returned to the Principal Component Analysis menu), select Initial eigenvalues and Scree plot.

Initial Eigenvalues

Component	Eigenvalue	% of Variance	Cumulative %
1	4.8029	53.366	53.4
2	2.5112	27.902	81.3
3	0.7331	8.145	89.4
4	0.3514	3.904	93.3
5	0.2072	2.302	95.6
6	0.1687	1.874	97.5
7	0.1249	1.387	98.9
8	0.0760	0.844	99.7
9	0.0247	0.274	100.0

The Kaiser criterion is to consider only components with eigenvalues above 1. In this case, only two acceptable components can be clearly discerned.

Scree Plot

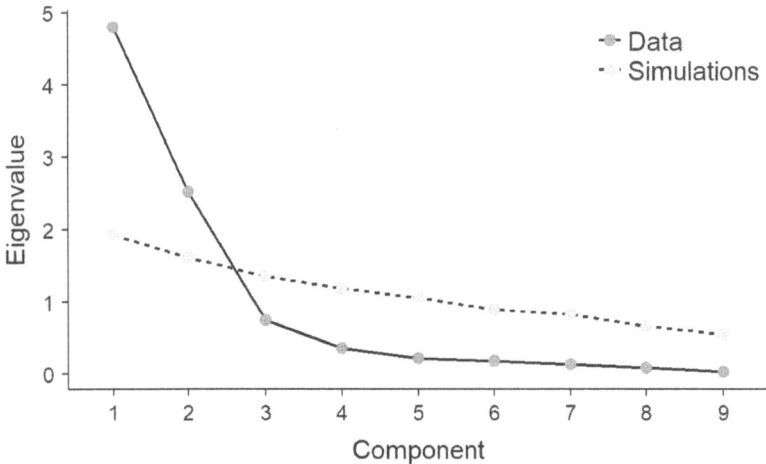

This is the scree test, or scree plot. The line of interest to you and me is the dark line resembling a cliff edge; towards the bottom the cliff comes to an end and what is left is a lot of scree, small loose stones. This plots each component number against its eigenvalue. As we already know, component 1 has an eigenvalue of 4.8, component 2 with eigenvalue 2.5 and so on.

What the user has to do with the scree plot is to decide where the cliff's descent finishes and the scree starts. The number of marks before the scree represents the number of valid components. In this case, the choice of three components seems to be fairly clear.

Both criteria have critics. As you might guess, the scree test becomes more difficult with more components, as deciding where the cliff ends and the 'shoulder' of the scree begins becomes much more subjective. The Kaiser criterion is considered to inflate the number of variables somewhat (Kaiser came to this conclusion himself).

Note that if you returned our dodgier variables, C, D and E, you will see that the Kaiser criteria moves to a three-factor recommendation, as the third component's eigenvalue swells up to more than 1. Both the Kaiser rule and the scree test are less robust than parallel analysis when the sample is small.

Do not dismiss these methods entirely. When the size of the sample runs into the hundreds, parallel analysis becomes less useful as a way of working out the ideal number of components (Revelle 2016), but the scree test is much more efficient with larger samples.

The Eigenvalues option gives you the opportunity of viewing the data differently. The default setting of 1 of course follows the traditional Kaiser criterion.

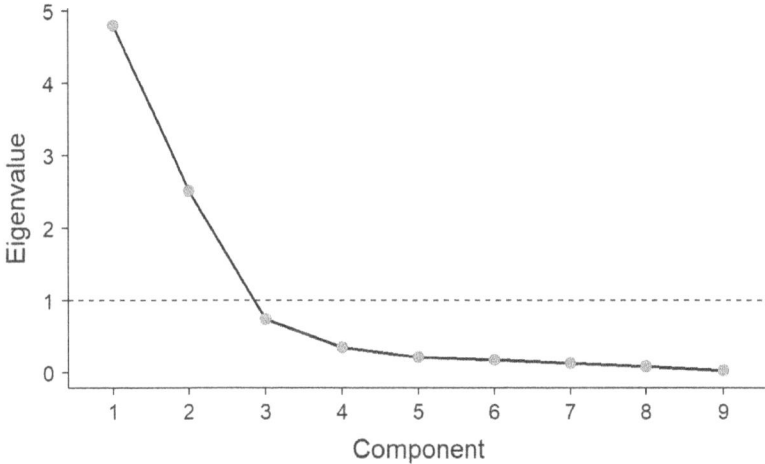

Note the horizontal plot line. The number of components above the line count; here, the Kaiser criterion shows two factors.

Summary

Component	SS Loadings	% of Variance	Cumulative %
1	4.65	51.7	51.7
2	2.66	29.6	81.3

The summary statistics show that together the biggest components take up over 80% of the variance. Let's use the Fixed number option and see what four options would look like:

● Fixed number

4   component(s)

Component Loadings

	Component				
	1	2	3	4	Uniqueness
A		0.861		0.372	0.0512
B		0.934			0.1081
F		-0.864			0.0424
G	0.385		-0.883		0.0493
H	0.881				0.1138
I	0.827	0.327	-0.374		0.0650
J	0.919			-0.324	0.0393
K	-0.837		0.424		0.0336
L	-0.864		0.362		0.0987

Note. 'varimax' rotation was used

I selected four to show two signs of instability within components. Component number 4 has only two variables. Component number 3 tends to share its variables with other components.

The component summary is quite instructive:

Summary

Component	SS Loadings	% of Variance	Cumulative %
1	3.991	44.35	44.3
2	2.624	29.16	73.5
3	1.423	15.81	89.3
4	0.360	4.00	93.3

While component 3 might be useful, given the size of its variance, component 4 is clearly negligible in terms of its influence. Although just how valuable the third component is probably a matter of judgement at the end of the day: does it represent something meaningful or useful, or is it merely a mathematical artefact? That is for the researcher to decide. If you really want to be completely traditionalist, you could opt for the 'None' rotation choice, returning to Parallel Analysis or based on Eigenvalue = 1. It also offers a two-component solution, but it is less than clear.

Component Loadings

	Component		Uniqueness
	1	2	
A		0.873	0.1792
B		0.933	0.1167
F	−0.549	−0.710	0.1945
G	0.647	−0.398	0.4225
H	0.810	−0.424	0.1646
I	0.960		0.0695
J	0.781		0.3701
K	−0.965		0.0684
L	−0.945		0.1005

Note. 'none' rotation was used

This higgledy-piggledy arrangement is a feature of unrotated PCA. In order to get a view of how the variables are interrelated, you would definitely need to visit a correlation matrix and at this stage your only clue would be the number of components indicated by the program. That is why we generally use rotation in order to get a clearer view, formally known as simple structure.

So it looks likely that we have a two-component model. On the other hand, we cannot rule out a third component altogether. Indeed, with a large sample, I would be inclined to follow the advice of the scree plot and use the Fixed number option as appropriate.

Part of the work is having a rationale; how well do the findings reflect on the context of your study? In the real world, we would probably have quite a good idea of whether or not a third component is likely to have explanatory power. We are likely to understand the relationship between at least some of the variables, giving us more of a clue as to what a factor consists of and its relative importance. Real world considerations usually inform us whether or not smaller factors have any meaning or are merely mathematical constructs.

Before leaving the PCA part of the program, I would like to point out the fact that the Based on eigenvalue setting can be overridden. If you wish to raise the difficulty of finding factors, perhaps because a large data set is throwing up too many, you can make the number bigger. If you wanted to increase the number of factors, maybe because you felt that the current setting didn't reflect the relationships between questions that you felt were really there, you could reduce the Eigenvalue to a smaller number such as .7, thus lowering the threshold for acceptable components and creating more.

Exploratory factor analysis

We will look at a new data set, Recruiter.csv (Kendall 1975). We are interested in how a recruiter has made assumptions about 48 potential employees, looking at two rival models. A two-factor model expects two likely groupings, one involving personal impressions such as soft skills and potential, the other being experience. The three-factor model adds sociability to the previous two factors.

Open Recruiter.csv and select the Factor menu, choosing the Exploratory Factor Analysis tab from the drop-down menu. Use all of the variables except 'Obs', the case numbering variable.

All of your favorite rotation flavors from PCA are here, with the same options, except that Rotation is set by default to Oblimin, an oblique rotation, and that there are model fit measures in the Additional Output.

As we are imposing our own ideas of what the models should look like, that is, either a two- or three-factor, I use the fixed number of components. An obvious point, by the way: when comparing models with different numbers of factors, you should compare like with like, using the same settings when reporting on each model. So here, both models will be examined using the Oblimin rotation.

Factor Loadings

	Factor		
	1	2	Uniqueness
Form of letter of application		0.629	0.546
Appearance	0.516		0.708
Academic ability			0.954
Likeability	0.574		0.666
Self-confidence	0.930		0.230
Lucidity	0.900		0.223
Honesty	0.583	−0.352	0.674
Salesmanship	0.851		0.215
Experience		0.865	0.288
Drive	0.757		0.273
Ambition	0.865		0.238
Grasp	0.863		0.176
Potential	0.829		0.169
Keeness to join	0.643		0.548
Suitability		0.757	0.216

Note. 'oblimin' rotation was used

The Factor Analysis program refers to each column as a Factor. On the whole, the results we see here seem to fit with my notions of the two-factor model, with loadings generally based on either impressions or experience.

Factor 1 groups together impressions of employees' soft skills with their potential. It is noticeable that all of the variables have positive factor loadings; negative loadings are possible (a measurement of social clumsiness, for example, would probably appear as a negative). Note that the factor is judged by the relative sizes of the loadings: high loadings suggest stronger factor contributions by those variables. This implies that Appearance, Honesty and Likeability, with factor loadings of less than 0.5, might not be useful, comparatively speaking. We can conclude that the recruiter might see some potential in people with good looks who appear honest and likeable, but may still prioritize qualities such as ambition, drive, grasp, lucidity, salesmanship and self-confidence.

Factor 2 particularly groups together impressions of suitability, experience and the form of the letter of application. The suitability of the candidate is probably associated with experience, perhaps with the form of the letter reflecting this experience. (While this may on the face of it seem like a sensible judgement, do note that this favors people who already have experience and may therefore entrench privilege, also depriving the employer of potentially more effective employees.)

Model Fit Measures

	RMSEA 90% CI				Model Test		
RMSEA	Lower	Upper	TLI	BIC	χ^2	df	p
0.178	0.149	0.213	0.689	−101	193	76	< .001

This table can be selected from the Additional Output options (Model fit measures). The RMSEA and TLI statistics are intended to reflect the fit of the particular model. RMSEA (Root Mean Square Error of Approximation, the "Ramsey" to its fan club) is considered to be a useful statistic. The Tucker Lewis Index (otherwise known as the Non-normed Fit Index or NNFI) is another commonly reported statistic.

To indicate a good fit, the RMSEA should be as small as possible and the TLI should be high. (Both usually improve after dodgy variables have been jettisoned.) Both should be reported, along with the RMSEA's confidence intervals.

As there can be some variation in these statistics because of sample size, I would report the sample size as well. In fact, it would be no bad thing to mention the number of variables in the same table. Although cut-off values for acceptable models are mentioned in the literature, I am less than happy about using them. As well as the obvious point about sample variation, a good model fit is not necessarily the same as a well-specified model (Kenny 2015).

The BIC (Bayes Information Criterion, or Schwarz criterion) should not be cited on its own. Its primary use is an easy way to compare models. A model with a smaller BIC is considered a better fit than one with a larger BIC. While models with more factors tend to get better (smaller) BICs reflecting their greater accuracy, the name of the game is to find out if the difference is great or not. If there is only a small difference, then you would probably prefer the more parsimonious explanation, incorporating less factors.

To find the difference: you merely subtract the smaller value from the bigger value (BigBic − SmallBic) and then report. The reporting of the difference is then:

0 to 2	Not worth more than a bare mention
2 to 6	Positive
6 to 10	Strong
> 10	Very Strong

Do note that the BIC can be affected by the inclusion or addition of variables. To use the statistics properly, only vary the models by the number of factors in use. (If you have removed variable Honesty, for example, this must be recorded and must remain the same when comparing BIC statistics.)

We will apply the BIC to our two-factor versus three-factor example shortly, but I would mention that the BIC statistic is not without its critics. On top of their technical criticisms, Gelman and Rubin (1995) "believe model selection to be relatively unimportant compared to the task of constructing realistic models that agree with both theory and data." (Expect a little more on similar themes at the end of the chapter.)

Before moving on to the three-factor model, we could consider the automated methods of choosing the number of factors. As this does not really fit with the theory-driven method, I'll show no images (you have seen them in the PCA section). The scree plot comes down firmly in favor of two factors, as do the Kaiser criterion and parallel analysis.

Number of Factors

○ Based on parallel analysis

○ Based on eigenvalue

　Eigenvalues greater than　0

◉ Fixed number

　3　factor(s)

Let us now examine the three-factor solution, as applied by the user.

Factor Loadings

	Factor			
	1	2	3	Uniqueness
Form of letter of application		0.669		0.492
Appearance	0.394			0.708
Academic ability				0.931
Likeability			0.955	0.100
Self-confidence	0.989			0.136
Lucidity	0.843			0.215
Honesty		−0.346	0.693	0.410
Salesmanship	0.916			0.142
Experience		0.838		0.323
Drive	0.730			0.262
Ambition	0.942			0.152
Grasp	0.773			0.180
Potential	0.679			0.179
Keeness to join			0.598	0.380
Suitability		0.762		0.225

Note. 'oblimin' rotation was used

The third factor has three large loadings. The one shared variable is only lightly loaded on the other factor. The combination of Likeability, Honesty and Keenness to join provides quite an interesting insight into the possible assumptions of the recruiter.

Factor 3 probably represents how the recruiter sees employees as being 'easy to get along with'. Sociability does seem to exist, if I have chosen the right term, in that the person is seen to be likeable and honest; perhaps this answers the question of "will he fit in well?" For some recruiters, the issue of 'likeability' is becoming a bigger factor for success at work, affecting how people are assessed by managers and treated by co-workers. Honesty is also important; put negatively, perceived dishonesty can lead to conflicts within teams and uncertainty in assessment. Keenness to join perhaps suggests an affinity with organizational norms.

It is also possible that the results shown above could lead to indications of preferred roles within an organization. Strength on the first factor – salesmanship, lucidity and the rest – might be particularly suitable for sales. Employees who are seen as strong on the third factor may be suited to public relations.

You will notice that the variable Academic ability does not appear to be loaded on any of the three components. It could be posited, as Kendall (1975) suggests, that a fourth factor exists which centers on academic ability, with a negative loading on keenness. You may want to try fixing four factors, perhaps trying the varimax rotation to see if it gives a slightly different way of looking at the fourth factor.

Model Fit Measures

	RMSEA 90% CI				Model Test		
RMSEA	Lower	Upper	TLI	BIC	χ^2	df	p
0.150	0.116	0.189	0.774	−112	132	63	< .001

Returning to the 3-factor model (and using Oblimin), we see the model fit measures. While the RMSEA and TLI can also be cited, my main interest here is the BIC statistic. The two-factor statistic was -101. Note that we are looking at negatives here, so -112 is the smaller number, to be viewed more favorably. But we need to calculate the difference. If the difference is negligible, then the extra information from an additional factor may not be parsimonious, perhaps a case of overfitting (the model as a snapshot of the data, so accurate that it is unlikely to be replicable). As they are both negative numbers, we may dispense with the signs: $112 - 101 = 11$, which our reporting table indicates as being a very strong difference in favor of the three-factor model.

Returning to more general usage, it is always better to prioritize interpretability over a low BIC. Whether or not the fourth factor would be of use is a matter of interpretation. Do note that it is generally seen as acceptable to try out a different rotation such as varimax to see if that provides a fresh insight into the data (Kendall 1975).

Controversies

Controversy 1 – Deciding on component and factor numbers

You will already have seen differences of opinion within the realm of statistics bubbling away, p values and the use of non-parametric tests being the ones that come easily to mind. While probably nothing matches the fierceness surrounding much of the debate between classical and Bayesian statistics, mentioned at the end of the book, factor analysis is notable for the sheer proliferation of disagreement; just pick a card, any card...

This first controversy is probably a matter of history, technology and common usage more than a source of heated debate. If we leave aside the theoretical ambiguity of the practice of selecting numbers of factors before conducting a factor analysis, the question is about the choice between the three methods shown in this chapter.

The Guttman-Kaiser criterion is commonly used in much of the traditional research literature. Kaiser found that the criterion usually named after him tended to result in a high proportion of components to variables. Not only that, but the proportion also tended to vary. These days, it is commonly considered to inflate the number of components. Choosing the Eigenvalues option is an acceptable way of trying to gain insights, but most researchers today agree that the Kaiser criterion should not be used by itself.

Parallel analysis as an automatic arbiter is very good for smaller samples. However, with observations running into the hundreds or more, Revelle (2016) and I have found that PA is likely to produce lots of components/factors with singleton or dual variables.

Fortunately, the scree test is much more sensitive when applied to larger data sets. So use the Parallel Analysis option for smaller data sets and the scree test for data sets running into the hundreds and greater. Use the Kaiser criterion as an additional tool for potential insights.

Let's share the end of this technical debate between the creator of varimax, the inventive Henry Felix Kaiser, and of parallel analysis respectively (see, statisticians can get along!):

> "The solution to the burning question of the correct number of common factors in a matrix of correlations is easy. Henry Kaiser used to say 'it was so simple he solved it every day before breakfast'." (Horn 1977).

Controversy 2 – PCA versus EFA techniques for factor analysis

If the previous debate was primarily a matter of the lag between increasing evidence and practice, this one is the source of much heat (and some light) between well-informed practitioners. You may well have noticed that I opened the chapter gingerly with my use of the terms data reduction and factor analysis. My use of the term 'factor analysis' covers a wide range of approaches and techniques, to indicate a difference between this and other areas of statistics, but some purists will object to even this usage, considering PCA to be a completely different field, theoretically and technically. To them, while questionnaire condensation, an apparently idea-free procedure, should be for principal components analysis, the search for meaningful factors should be a matter for factor analysis, and never the twain shall meet.

Yet there are researchers who advocate the use of PCA for both purposes. For example, Stevens (2009) claims that both techniques often offer similar results. While PCA may not be theoretically based, it can be used to similar effect. 'Wherefore meaningless?' as Shakespeare's Edmund did not quite say.

Cliff (1987) considers the dispute to be largely ideological, with some authorities viewing PCA as the only suitable approach, as FA methods "just superimpose a lot of extraneous mumbo-jumbo" related to factors that are basically impossible to measure. He considers exponents of factor analysis to hold even stronger views, with PCA seen as

> "at best a common factor analysis with some error added and at worst an unrecognizable hodgepodge of things from which nothing can be determined. Some even insist that the term 'factor analysis' must not be used when a components analysis is performed."

Borrowing selectively (and wickedly) from Pedhazur and Schmelkin (1991), I would note that various FA advocates recommend trying out various solutions, citing the importance of theory. "Not surprisingly, critics of FA view the wide latitude researchers have in choosing a factor-analytic solution as a prescription for unfettered subjectivity." They cite Reyment, Blackith and Campbell (1984) who consider FA's popularity to derive from the ability of researchers to impose their preconceived ideas on the raw data.

Stevens (2009) points out that both techniques often produce similar results and prefers principal components analysis for four reasons: Firstly, "it is a psychometrically sound procedure..." Secondly, he declares it to be mathematically simple. Thirdly, there is the problem of factor indeterminacy: theoretical factors are not clearly delineated and are difficult to distinguish from mathematical artefacts. Stevens cites Steiger (1979):

> "My opinion is that indeterminacy and related problems of the factor model counterbalance the model's theoretical advantages, and that the elevated status of the common factor

model (relative to, say, components analysis) is largely undeserved."

Stevens' fourth point is the need for an awful amount of reading for a proper understanding of factor analysis the technique.

For a start to your reading, and to help you come to your conclusions about this debate, I recommend any two of five readings in particular. For an in-depth description of these approaches, without opinions but covering quite a lot of technical issues, see Bryant and Yarnold (1995). For balanced descriptions, including the traditional separation of FA and PCA, but without judgements being made, try the relevant chapters in Tabachnick and Fidell (2007) or Pedhazur and Schmelkin (1991). For an online article describing the practice of EFA which comes down strongly in favor of factor analysis, see Costello and Osborne (2005). For an opinion favoring principal components analysis, see Stevens (2009).

One practical thing that you can try immediately is to try out the Recruiter.csv data with the Principal Components Analysis submenu. To simplify the loadings output, set 'Hide loadings below' to 0.5. Then try the traditional 'Based on eigenvalue' setting, using the default setting of 1 and see what you think of the factors indicated by PCA.

Controversy 3 – rotation methods

Although nowhere near as vociferously argued over as the previous issue, rotation is an area of constant debate and you will find contradictions and uncertainty in these waters. One source of relief stems from this: it is rare for your choice to be theoretically wrong!

The fundamental choice is between **orthogonal rotation** and **oblique rotation**. The orthogonal rotation methods separate the components or factors. Oblique rotation methods assume relationships between the factors.

This was why the default for PCA was an orthogonal rotation. If you carry out a straightforward principal components analysis, for example whittling away a number of questions by getting rid of similar ones, you assume that the components are going to be quite different from each other and anticipate this by pulling them apart.

In exploratory factor analysis, the choice can depend upon our intentions. If we are trying to uncover conflicting or distinct points of view about a social issue, we would use an orthogonal rotation method to create uncorrelated structures. If we are trying to see how disparate points of view might converge in some way, then an oblique method would make sense, allowing correlated structures (Costello and Osborne 2005).

Alternatively, we can take a data-driven perspective, using rotations that reflect the nature of the relationships between the variables. Start with an oblique rotation and then see if the factors are reflected in the correlations of the variables matrix (use the Correlations matrix). If they are, judging by correlations of greater than .32 (or smaller than −.32), then continue with oblique. If the structures do not match up with the correlations, move to orthogonal rotations (Tabachnick and Fidell 2007).

In terms of how they are used in the field, orthogonal rotation is most commonly used, possibly because the results are easier to interpret. On the other hand, oblique rotation is often considered more theoretically realistic by traditionalists using factor analytical techniques.

In allowing correlated structures, oblique methods can be said to be more accurate, hence their popularity with traditionalist users of factor analysis as a set of techniques. However, this does make it possible for **overfitting** of the model in question. This means that the model can create almost a snapshot of the actual data, very accurate but too complex, hugging the data like a set of over-tight jeans, so that there may be problems replicating the model with fresh data. (Under-fitting means that there is too much 'noise' – the jeans are hanging down way too far.) Orthogonal methods produce less accurate models, having less good a fit, but such models are probably more parsimonious and more likely to be replicable.

Note: If the loadings tend to be small, you will often find quite similar results from both sets of rotation techniques. Similarly, if the correlations between the factors are small, then they are likely to be similar (Rennie 1997). For the latter, select the Factor correlations output option.

Orthogonal rotation techniques

Of the orthogonal rotation techniques, varimax, Kaiser's brainchild, is the most commonly used. It creates factors which have high correlations with one subset of variables, and low or no correlations with another subset (Stevens 1996). Although generally a good choice for a simple interpretation of factors (Rennie 1997), it is "inappropriate if the theoretical expectation suggests a general factor" (Gorsuch 1983). A general factor is something like intelligence (g) where everything loads onto a single factor for a good theoretical reason.

The quartimax rotation is probably the most suitable orthogonal rotation for a general factor (Pedahazur and Schmelkin 1991). This cleans up the variables (Stevens 1996) by forcing each variable to correlate on one factor while having little or no relationship with other factors. This makes variables easier to interpret (Rennie 1997).

Oblique rotation techniques

Of the oblique methods, oblimin is probably the best-known, 'classic', rotation and a common default amongst researchers. The promax rotation method is quicker than oblimin and thus useful with large data sets, but has another advantage: although an oblique rotation, promax was adapted from an orthogonal rotation and is more likely to produce replicable results (Rennie 1997). A proponent of simplimax claims that "this method can recover relatively complex simple structures where other well-known simple structure rotation techniques fail" (Kiers 1994).

General points about rotation

I have cited the different rotation methods as if each were designed for a purpose. I would like to make two points which rather qualify this.

Firstly, the results of different rotation methods are often not hugely different. Personally, I find this quite reassuring, although perhaps less than exciting. On the other hand, as I sometimes notice using varimax, something may come up..

Secondly, theorists and empiricists usually agree on the idea of using several rotation methods in the search for simple structure. Both combatants share the desire to seek a clearer view of what are essentially the same phenomena.

Beyond the technicalities

I conclude this chapter with some notes, a few of which might accord with what you have already seen. The number of factors is the most important issue in an exploratory factor analysis, more so than such issues as the choice of rotation. Although it is true that more factors give you a better fit with the data, less factors offer a more parsimonious explanation; a trade-off is usually sensible (Tabachnick and Fidell 2007).

There are difficulties in both PCA and FA, but "a good PCA or FA makes sense. A bad one does not" (Tabachnick and Fidell 2007). If you are reducing the number of variables, fine. If you are exploring a theory, also fine. But,

> "exploratory FA is not, or should not be, a blind process in which all manner of variables or items are thrown into a factor-analytic 'grinder' in the expectation that something meaningful will emerge. .. GIGO ... garbage in, garbage out." Pedazhur and Schmelkin (1991).

If, for example, you throw in scales of very different complexity, it is quite likely that they will gather together in factors, dominating the rest of the analysis. Pedazhur and Schmelkin also note that,

"... FA is not a method for uncovering real dimensions, or reality underlying a set of indicators. .. almost anything can be uncovered, if one tries long and hard enough e.g., extracting different numbers of factors, using different methods of rotation of factors... meaningful application of FA is unthinkable without theory. .. When you have no theoretical rationale for doing a FA, DON'T!"

Chapter 14 – Logistic regression

Are tennis players from particular backgrounds more likely to drop out of professional tennis than people who are better off? Are married women who do not have children more likely to succeed as sports coaches? Are there particular sets of attributes which mean that some people are more likely to take illegal performance-enhancing drugs than others. Are different levels of motivation towards training influenced by race, gender or parental pressure?

This type of question can be answered by logistic regression, also referred to as logit regression or the logit model. In each case, there are specific outcomes, or 'outputs', which are categorical. So the logic of most of the previous tests has been reversed. ANOVA, *t* tests and correlations had a condition or grouping as the independent variable / predictor, with the dependent variable as a continuous or ordinal measure. Here, the dependent variable is categorical. Put another way, we are seeing if data fits into certain discrete categories.

Also of note is that logistic regression's independent variables can be categorical, ordinal or continuous, often in combination. This versatility makes logistic regression very popular among researchers. It is only the dependent variables, our outputs or classifiers if you like, which have to be categorical (although this can include ordered categories).

There are three forms of logistic regression, depending upon the nature of the dependent variables (outputs): binomial regression, multinomial regression and ordinal regression. Binomial, also known as binary regression, has a simple dichotomous output: two categories only. Multinomial regression has more than two discrete outcomes. Ordinal regression has not only more than two outcomes, but these are ordered.

As an example of where the different tests could be used, let us consider predicting the likelihood of a team winning or losing. We would use binomial regression to see if certain attitudes or attributes indicated a likelihood of a team winning or losing. If draws are quite common in the chosen sport, we may wish to add the draw as a third output category, so multinomial logistical regression would be suitable. If we really wanted to complicate our lives (and maybe put our finances at risk), we may consider multiple bandings such as 'close call', 'clear win' and 'walkover', using ordinal logistic regression; the point here is that the multiple outputs are hierarchical.

Further binomial logistic regression examples could be yes or no to an issue, recruitment to a senior management position or not, likely success or failure on a course, or whether or not a technique successfully deals with a problem.

Multinomial regression, rather intuitively, means that you have more than two discrete outcomes. Less intuitive are some of the alternative names for this method, such as polytonous logistic regression and maximum entropy. Possible extensions of the above examples would be yes/no/don't know; senior/supervisory/standard; dropped out/graduated /failed; complete success/partial/failure.

Ordinal regression is usable for ordered outcome categories (aka ranked or hierarchical). Senior/supervisory/standard could perhaps be better placed here, where a hierarchy is assumed. Other outputs could be: excellent/very good/good/fair/poor; gold/silver/bronze; big/medium/small; likely/unlikely/outside chance. It is possible to use multinomial regression with hierarchical categories, but you would lose information that way.

Assumptions

One reason logistic regression has only recently entered introductory textbooks is because its iterative nature made it difficult to use if you didn't have access to a fast computer (I once spent all night waiting for calculations not to work). Logistic regression is particularly popular because it can handle a combination of independent variables of different types; a person's scores on Likert scales and examinations can be combined with their gender and their employment status all in the same calculation. Its relatively undemanding assumptions also make it more usable in many contexts than its nearest rival, discriminant analysis, which is MANOVA in reverse and shares that test's assumptions for data.

The assumptions for logistic regression are fairly easy to meet. Groups need to be exclusive and exhaustive: each case can only fit into one dependent variable category and all cases must be allocated. Statistical independence is required: each case must occur only once.

Another assumption is a linear relationship between continuous independent variables and the logit transformation of the dependent variable. If you are using continuous independent variables, a possible violation of the assumption means that logistic regression may underestimate the strength of an effect. If a likely hypothesis has been rejected, a practical solution is to categorize the continuous variables – in a meaningful way – to see if this has any effect on the outcome. *

There is also the question of sample size. Wright (1995) cites Aldrich and Nelson (1984) in recommending 50 cases per predictor, also noting that it is sensible to be able to split a sample so that findings can be cross-validated with fresh data. Some people suggest 10 outcome events per predictor variable (EPV), although one study based on simulations indicated that this is too conservative for analytical purposes, although noting that greater strictness is required for the prediction of classification and that larger samples are always desirable (Vittinghoff and McCulloch 2007).

---

*For aficionados: I realise that the Box-Tidwell test is suitable for checking for violations, but this doesn't always work properly, so for intermediate users, I recommend categorization of continuous variables whenever this happens.

Suitable data set structures

The following data set contains only a few cases, just to demonstrate the structure for the three different types of logistic regression. I plead the same excuse for the rather crude sociological categories used.

The predictors in logistic regression data sets can be continuous variables, categorical variables, ordinal variables or any combination thereof. Logistic regression's popularity is very much based on its ability to absorb a range of data. In the following data set, Gender is a dichotomous (yes/no) categorical variable, as is Marital status. Area is categorical, while Class is categorical but could be considered as ordinal. Score and Parental are continuous.

Case	Outcome	Gender	Marital	Class	Score	Parental	Area
1	Graduated	Female	Unmarried	Upper	80	40,000	Inner
2	Graduated	Female	Unmarried	Middle	60	30,000	Suburb
3	Dropped	Female	Unmarried	Lower	50	26,000	Inner
4	Dropped	Female	Married	Middle	60	23,000	Rural
5	Graduated	Female	Married	Middle	50	30,000	Inner
6	Dropped	Female	Married	Lower	57	30,000	Inner
7	Graduated	Male	Unmarried	Upper	80	40,000	Suburb
8	Graduated	Male	Unmarried	Middle	75	30,000	Rural
9	Dropped	Male	Unmarried	Middle	50	32,000	Suburb
10	Graduated	Male	Married	Middle	70	37,000	Suburb
11	Dropped	Male	Married	Lower	58	28,000	Suburb

Let's say that graduation or otherwise is the dependent variable: I have called the classification variable Outcome. * This is dichotomous – students have either dropped out or graduated – and therefore binomial logistic regression would be used. However, in their current configurations as dichotomous variables, Gender or Marital could also be used as binary regression outcomes. If Area was the outcome, then multinomial regression would be necessary. Ordinal regression would be likely for Class; multinomial regression could be used, but we would lose the information available from the ordinal values.

---

*You can find examples in statistical literature of the outcome variable being referred to as 'Class', but this is not mandatory and for sociologists would be an obvious waste of a predictor name.

Chapter 14 – Logistic regression

Binomial logistic regression

The Housing.csv file contains a real data set, from a study by Wilner et al (1955). They were interested in the contact hypothesis, the idea that relationships between conflicting groups of people are more likely to improve with greater interaction. This was an early study of racially integrated versus segregated public housing. If the contact hypothesis is correct, then favorable feelings are more likely to accrue because of more frequent contact, mediated by proximity and people's expectations through social norms.[*]

Sentiment	Proximity	Contact	Norms
favorable	close	frequent	favorable
unfavorable	distant	infrequent	unfavorable

Each variable here has two levels only, close versus distant, frequent versus infrequent, and so on. Do note, however, that only the output variable, in this case Sentiment, actually needs to be dichotomous. The predictors could be multilevel, continuous or ordinal, in any combination.

---

[*]Peculiarly but perhaps of its time, white people were asked for their views, but not African-Americans (Pickren and Rutherford 2010).

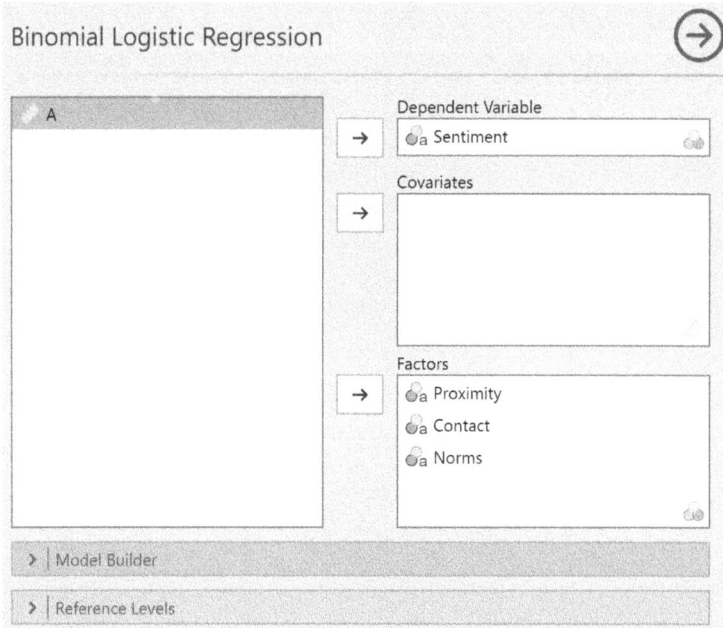

Use the Housing.csv file and, from the Regression tab, select 2 Outcomes / Binomial. The output variable, sometimes referred to as a 'classifier', is placed in the Dependent Variable box, with other categorical variables going into the Factors box. Any numerical data would fit into 'Covariates'. Ignore the "A" variable; this is a case variable.

Releveling as preparation for logistic regression

Before considering the results, it is recommended that you relevel the variables. This clarifies the direction of the results, as reflected in the coefficients. One condition ('level') in any given variable is used in the calculations as a baseline, also known as the reference, against which the other conditions are compared. Generally, you set the baseline to the level that you are not interested in, but this can be reversed, assuming that only one level is of interest and therefore you want to see the relationships with everything else.

Variable	Reference Level	
🔖 Sentiment	favourable	▼
🔖 Proximity	close	▼
🔖 Contact	frequent	▼
🔖 Norms	favourable	▼

Please open the Reference Levels section. The default reference for each variable is the first name in alphabetical order.

Variable	Reference Level	
🔖 Sentiment	unfavourable	▼
🔖 Proximity	distant	▼
🔖 Contact	infrequent	▼
🔖 Norms	unfavourable	▼

In this example, I want to consider the positive aspects of each condition, so I'll reference the negatives, reversing the lot. Sentiment is referenced as unfavorable, Contact is infrequent and so on. Releveling does have an effect. In the table to follow, all of the coefficients ('Estimates') for the predictors are positive. If, for example, you decided to reverse the levels for Sentiment, you would find them turning negative, indicating the opposite causal direction; if you changed, say, Contact to Frequent, then there would be a more localized alteration.

Basic reporting

Model Coefficients

Predictor	Estimate	SE	Z	p
Intercept	−1.0286	0.147	−6.983	< .001
Proximity:				
close − distant	0.0982	0.184	0.532	0.595
Contact:				
frequent − infrequent	0.9458	0.181	5.222	< .001
Norms:				
favourable − unfavourable	0.7960	0.177	4.506	< .001

Note. Estimates represent the log odds of "Sentiment = favourable" vs. "Sentiment = unfavourable"

The dependent (outcome) variable is Sentiment, whether or not people were positive about racially integrated housing. This is not referred to in the body of the table. The statistics in the main table would be reported.

The coefficients referred to as 'Estimate' are reported as *beta* (B). Their importance is in telling us the direction of the effect. If they are positive, then they indicate that a rise in the outcome, the level of Sentiment, is likely to occur as the predictor increases; in this example, more contact and positive norms mean more positive attitudes. Negative values point in the opposite direction. Scores near zero indicate no effect, as in the case of Proximity here.

Aside from the direction of the effect, the estimates also indicate the change in the log odds of the outcome variable for a one unit increase in the predictor variable. The log odds are a way of measuring probability, but more easily interpreted statistics, odds ratios, will be introduced shortly. Standard Error in your report appears here as SE. The z value is a popular measure; a score of 1.96 or greater usually means a probability of less than .05 (Wright 1995). The p value is based on z, which is the primary identifier of the significance of the variable in the model.

Model Fit Measures

Model	Deviance	AIC	R^2_{McF}
1	772	780	0.0805

The deviance and the Akaike information criterion (AIC) values help assess model fit which will be discussed further. Moreover, the AIC becomes important when we consider other models. If, for example, we had 5 variables and decided to try 4 instead, then we would compare the AIC of each model. A smaller AIC denotes a better fit. It should be noted, however, that there is often a trade-off between a parsimonious fit (using fewer variables) and the accuracy derived from more variables.

Model Coefficients

Predictor	Estimate	95% Confidence Interval		SE	Z	p
		Lower	Upper			
Intercept	−1.0286	−1.317	−0.740	0.147	−6.983	< .001
Proximity:						
close − distant	0.0982	−0.263	0.460	0.184	0.532	0.595
Contact:						
frequent − infrequent	0.9458	0.591	1.301	0.181	5.222	< .001
Norms:						
favourable − unfavourable	0.7960	0.450	1.142	0.177	4.506	< .001

Note. Estimates represent the log odds of "Sentiment = favourable" vs. "Sentiment = unfavourable"

You may also choose to report confidence intervals with the coefficients, an option available in the Model Coefficients section.

Before we go too far with these excitingly significant variables, we cannot at this point be sure that the overall model is efficacious. We need to consider the null hypothesis, that the predictors make no difference.

Omnibus Likelihood Ratio Tests

Predictor	χ^2	df	p
Proximity	0.283	1	0.595
Contact	27.757	1	< .001
Norms	20.461	1	< .001

An omnibus test is one which looks at the whole model (like the old-fashioned ANOVA before post hoc tests became acceptable). The likelihood ratio tests provide a simple way of examining the model.

This raises another reporting issue. Although we may well explore further models, in this case perhaps one with just the Contact and Norms factor, we would usually report a non-significant factor where the variable had theoretical value (in this case, I'll leave it to the sociologists to slug it out).

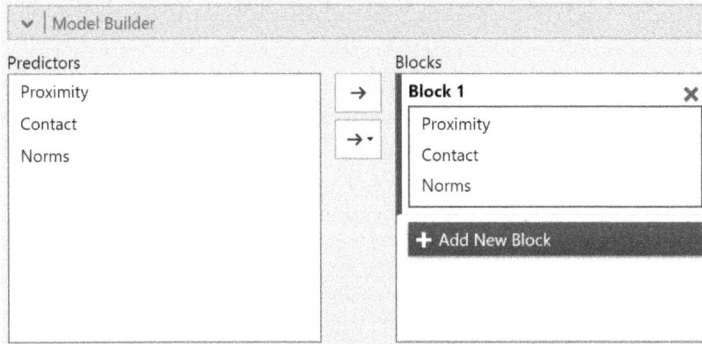

The default in the Model Builder is for all the variables to sit in the one block. Now let's compare a new model, omitting Proximity, with the original.

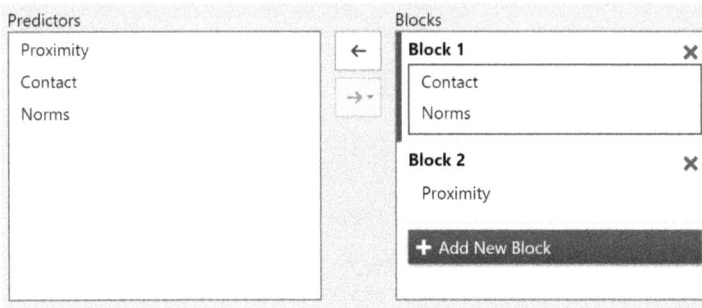

I have added a new block and shifted Proximity into it. Essentially, Block 1 is a new model, with just Contact and Norms. Block 1 and Block 2 together form the original.

Model Fit Measures

Model	Deviance	AIC	BIC	R^2_{McF}
1	773	779	792	0.0801
2	772	780	798	0.0805

Model 1 refers to the upper block, just containing Contact and Norms; Model 2 comprises both blocks together, in this case the original model.

A smaller AIC indicates a better fitted model, as is the case with BIC, which I have selected from the Model Fit section. Even if these were to be of the same size or a little bigger, this would not change my mind about preferring the more parsimonious model. It is to be expected that the model with more information would be more accurate and in a case where the difference is marginal, it would seem reasonable to adopt the model with less variables.

Model Comparisons

Comparison				
Model	Model	χ^2	df	p
1	- 2	0.283	1	0.595

The model comparisons table confirms me in my prejudices. There is clearly no significant difference between the models, with a very small Chi squared statistic and a large p value. So, the addition of Proximity makes little difference to the model. There is no reason why we should not accept the new model.

Interpreting the coefficients

Model Specific Results [Model 1 ▼]

Model Coefficients - Sentiment

Predictor	Estimate	SE	Z	p
Intercept	−1.008	0.142	−7.10	< .001
Contact:				
frequent − infrequent	0.973	0.174	5.58	< .001
Norms:				
favourable − unfavourable	0.810	0.175	4.64	< .001

Note. Estimates represent the log odds of "Sentiment = favourable" vs. "Sentiment = unfavourable"

We can toggle between reporting tables for the models. Here, the new, smaller, model is seen to have slightly larger coefficients than those of the original.

As noted before, the basic coefficients give you information about direction, but they are not otherwise particularly usable. What we can do is to use the odds ratios, to be found in the Model Coefficients section.

Model Specific Results Model 1 ▾

Model Coefficients - Sentiment

						95% Confidence Interval	
Predictor	Estimate	SE	Z	p	Odds ratio	Lower	Upper
Intercept	-1.008	0.142	-7.10	< .001	0.365	0.276	0.482
Contact:							
frequent – infrequent	0.973	0.174	5.58	< .001	2.645	1.880	3.721
Norms:							
favourable – unfavourable	0.810	0.175	4.64	< .001	2.248	1.596	3.166

Note. Estimates represent the log odds of "Sentiment = favourable" vs. "Sentiment = unfavourable"

What the odds ratio does is to tell us that for every unit increase of the predictor, the odds of the outcome variable will be influenced by the amount shown. In the above case, we are dealing with categorical predictors: frequent contact makes the odds of having a favorable sentiment about two and a half times more likely than infrequent contact. The confidence intervals are included to show that there is some inaccuracy inherent in such a claim.

If our Contact predictor had been a Likert scale, then each unit difference on the Likert scale would make the odds of having favorable sentiment about two and half times more likely (the same terms and conditions apply). If it had been a continuous measure, let's say in units of U.S. dollars, then each dollar would make the odds of having a favorable outcome about two and a half times more likely.

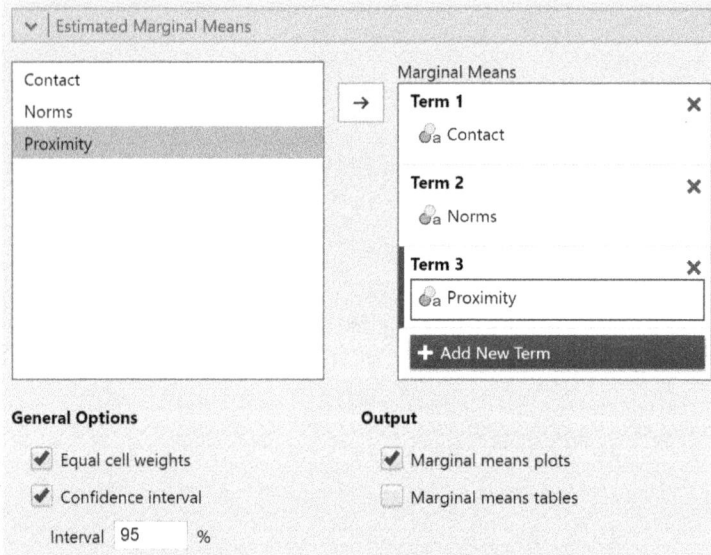

Now return to the full model (Model 2). A visual examination of the coefficients is available in the Estimated Marginal Means section. By placing each dimension in its own box, we can examine them separately. To study interactions between effects, put them in the same box.

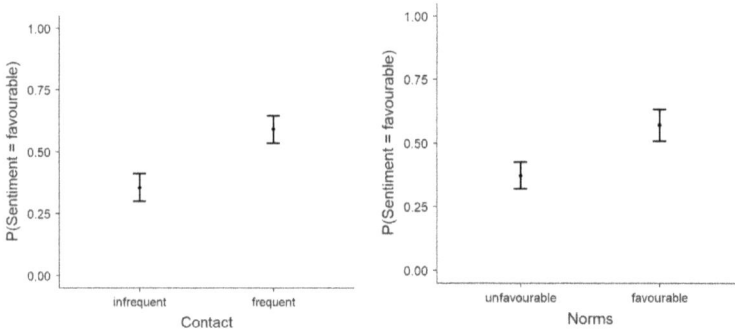

These graphs rather speak for themselves. [*]

---

[*]Estimated marginal means examine the dimensions within the context of their overall model, adjusting for the other factors. This means that interpretations using the same data can vary somewhat within different models.

Pseudo R-squared statistics

Model Fit Measures

Model	Deviance	AIC	BIC	R^2_{McF}	R^2_{CS}	R^2_N
1	773	779	792	0.0801	0.105	0.140
2	772	780	798	0.0805	0.105	0.141

One thing that does not exist in logistic regression is the type of R squared statistic that you have used previously for effect size. Here, we get three 'pseudo R-squared' statistics, selected from the Model Fit section. These are McFadden's R-squared, Cox and Snell's (aka Maximum Likelihood) R-squared and Nagalkerke's (aka Cragg and Uhler's) R-squared respectively. These are measures of strength of association, but unlike R-squared, they should not be used as estimates of variance. They are generally smaller than the usually reported statistics and their critics suggest that they should not be reported at all. Hosmer and Lemeshow (2000), logistic regression gurus, put this in perspective:

> "...low R2 values in logistic regression are the norm and this presents a problem when reporting their values to an audience accustomed to seeing linear regression values. ... we do not recommend routine publishing of R2 values with results from fitted logistic models. However, they may be helpful in the model building state as a statistic to evaluate competing models. "

When comparing different models, the comparison should be made against the same statistic, for example McFadden for Model 1 against McFadden for Model 2 (Veall and Zimmermann 1996). Each of these statistics is said to have its merits, although a fair amount of support these days is for the McFadden; a score of .2 to .4 is considered to reflect an excellent fit (McFadden 1979).

The great similarity in pseudo R-squared statistics between the models provides another argument in favor of our more parsimonious model, containing just the variables Contact and Norms.

Prediction

In many cases, when we are trying to achieve true classifications, we may want to make predictions. Are people with a particular profile reliably going to drop out of a course? or commit a serious crime? or become a leading diplomat?

For the sake of continuity, we will continue for a while with the study previously under consideration. Do note that while we have good results for the purposes of our study, it is highly unlikely that this is going to provide reliable classifications; thinking of real life, how likely is it that increased contact is going to convert almost everybody to tolerance and good will? or that low contact is always going to indicate racism in most of the population? So this part of the data is primarily to show how prediction is carried out; just don't expect a spectacular output.

Classification Table – Sentiment

Observed	Predicted unfavourable	favourable	% Correct
unfavourable	237	88	72.9
favourable	133	150	53.0

Note. The cut-off value is set to 0.5

Select all of the Predictive Measures from the Prediction section, and also the ROC options. As I suggested, accuracy is low, but the regression does predict over 60% of the cases. The Classification Table gives a breakdown of the actual figures of how well the model predicts the outcomes. The program did rather better in its categorization of unfavorable outcomes than it did with favorable outcomes

Predictive Measures

Accuracy	Specificity	Sensitivity	AUC
0.637	0.729	0.530	0.684

Note. The cut-off value is set to 0.5

The proportions to the right of the Classification Table are taken and used for the measures of Sensitivity and Specificity. Sensitivity is the proportion of observations in which the measure gets a positive result right; high sensitivity means getting a large proportion of true positives. Specificity is the proportion of observations in which the measure correctly indicates a negative; high specificity means getting a high proportion of true negatives correct.

ROC Curve

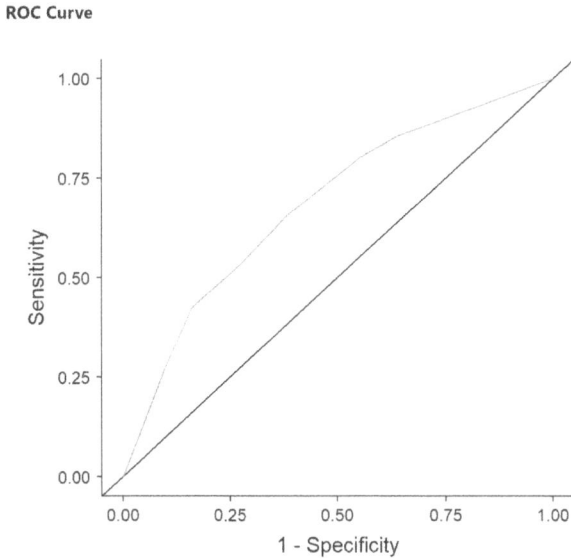

A traditional way of examining these issues is the ROC curve, * which shows how well a test or tool diagnoses a binary option. If this had been a diagnostic test, I would have wanted a great big mountain of a curve, taking up most of the space. The AUC, the area under the curve, is a measure of overall accuracy. The larger proportion of the chart's space under the curve, the better. If you select the AUC option, you will get

---

*Receiver Operating Characteristic. During World War Two, the US wanted to improve the accuracy of radar. They did this by measuring radar operators' ability to distinguish Japanese aircraft. More recently it has been used in diagnostic medicine.

0.684, a less than impressive statistic. The following is a guide to overall accuracy using the AUC:

- .90-1 = excellent (A) (1 = Perfect)

- .80-.90 = good (B)

- .70-.80 = fair (C)

- .60-.70 = poor (D)

- .50-.60 = fail (F) (worthless)   *

Our model's accuracy is poor. However, this is a study rather than a diagnostic test. We can use AUC as an approximate measure of accuracy. Approximately 60% accuracy is not actually bad when the context is considered; a study like this is likely to have a good deal of variance.

If we were looking for a balanced approach to sensitivity and specificity, we could look at the curve, looking for the point on the curve nearest to the top left corner. If my eyesight serves me well, the sensitivity is a bit above .6 and the specificity is rather under that figure.

Real-life problems usually require a balancing of priorities. If we take the example of trying to predict results, a test with an extremely high sensitivity may detect a lot of wins, but they may be accompanied by rather a lot of losses. On the other hand, a test with high specificity, that is, avoiding false negatives, means that we are unlikely to get any losses in our predictions, but we risk missing some victories too.

If you were trying to make operational decisions, you might want to consider thresholds. You may assign the threshold to optimize sensitivity, or specificity depending on which is most important in the situation.

---

*A negative would mean that the test is actually diagnosing in the wrong direction (Hajian-Tilaki 2013).

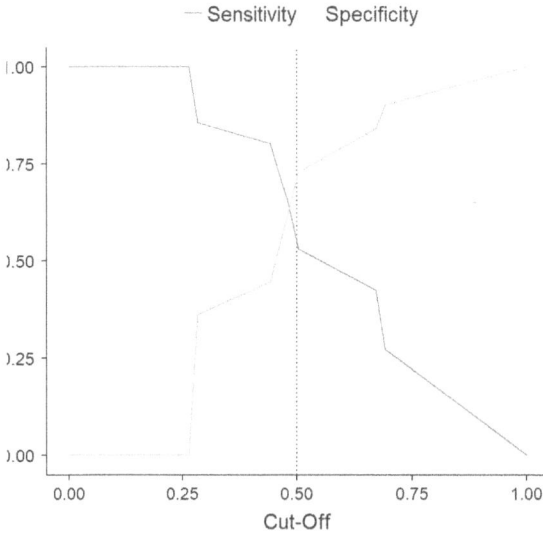

The Cut-off plot from the Prediction section shows the default 0.5 cut-off value. This can be changed to lower and higher cut-off values, moving the dotted line to the left or the right. A shift toward higher sensitivity would mean poorer specificity; higher specificity means poorer sensitivity. You would have to consider what measures could be taken to adjust to an acceptable threshold.

Multinomial logistic regression

Age	Education	WantsMore	Usage
< 25	High	No	No
25-29	Low	Yes	Yes
30-39			
40-49			

Married Fijian women who were able to have children were interviewed to find out about their level of education, their desire to have more children and whether or not they were using contraception at the time of interview (Little 1978). Age is our outcome variable; we want to see whether or not an age-based analysis of contraceptive use makes sense.

Open the Contr.csv file, press the Regression button and select N outcomes / Multinomial.

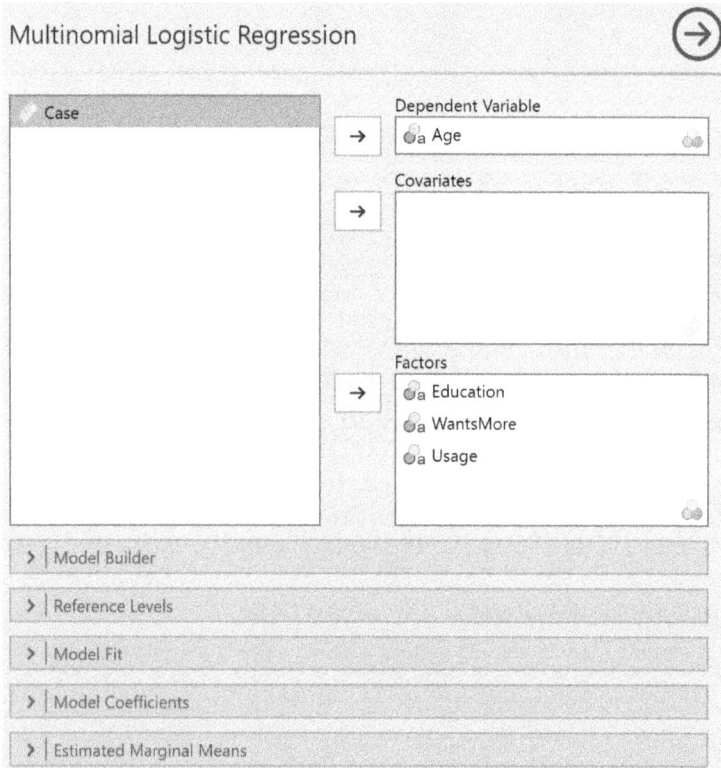

Multinomial Logistic Regression

Case

Dependent Variable

→ Age

Covariates

→

Factors

→ Education
    WantsMore
    Usage

> Model Builder

> Reference Levels

> Model Fit

> Model Coefficients

> Estimated Marginal Means

The dialog box is the same as for binomial regression. Numerical data would go in the Covariates box.

Reference Levels		
Variable	**Reference Level**	
🔒a Age	<25	▾
🔒a Education	High	▾
🔒a WantsMore	No	▾
🔒a Usage	No	▾

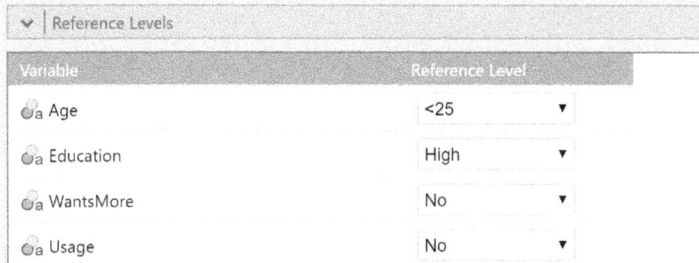

As previously, releveling should be considered. Here I select the under 25 Age level, which will be compared with the older age groups. On reflection, I think I'll leave the default baselines for the other variables.

Model Coefficients

Age	Predictor	Estimate	SE	Z	p
25-29 - <25	Intercept	0.264	0.163	1.62	0.106
	Education:				
	Low – High	0.444	0.174	2.55	0.011
	WantsMore:				
	Yes – No	-0.569	0.171	-3.34	< .001
	Usage:				
	Yes – No	0.411	0.176	2.34	0.019
30-39 - <25	Intercept	0.553	0.152	3.64	< .001
	Education:				
	Low – High	1.517	0.159	9.53	< .001
	WantsMore:				
	Yes – No	-1.293	0.159	-8.12	< .001
	Usage:				
	Yes – No	0.939	0.165	5.68	< .001
40-49 - <25	Intercept	-0.882	0.210	-4.20	< .001
	Education:				
	Low – High	2.364	0.214	11.03	< .001
	WantsMore:				
	Yes – No	-2.115	0.217	-9.76	< .001
	Usage:				
	Yes – No	1.184	0.215	5.50	< .001

To save unnecessary verbiage, I will assume that you have read the preceding section on binomial logistic regression. Essentially, positive and negative coefficients (beta) represent directions of effect for the predictors, with very near zero (not shown here) meaning no appreciative effect. Each block contains coefficients for variables pertaining to each outcome (the levels of the dependent variable). The first block compares Age = 25-29 to our baseline Age < 25, the second block compares Age = 30-39 to the baseline and likewise with the final block. As previously, you can also report confidence intervals for the coefficients by going into the Model Coefficients section.

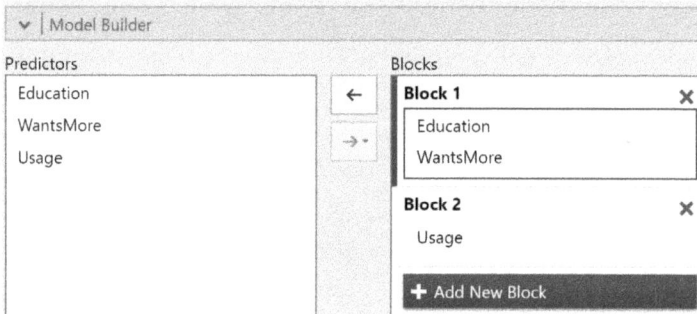

You can also see if a better model is to be had, for example by shifting Usage to its own block in Model Builder and looking at measures such as the AIC.

Model Fit Measures

Model	Deviance	AIC	BIC	R^2_{McF}
1	3846	3864	3913	0.0902
2	3802	3826	3890	0.1008

Model Comparisons

	Comparison			
Model	Model	χ^2	df	p
1	- 2	44.8	3	< .001

Model 1 is the same as Block 1 in the Model Builder. Model 2, the comprehensive model, has distinctly smaller AIC and BIC values, and there is a significant difference between the models, so the original, more comprehensive, model is to be preferred.

Omnibus Likelihood Ratio Tests

Predictor	χ^2	df	p
Education	190.7	3	< .001
WantsMore	129.1	3	< .001
Usage	44.8	3	< .001

If we choose the Likelihood ratio tests from the Model Coefficients section, not only do we see that each of the three variables is significant in relation to the age outcomes, but we have some gauge of their influence: a large likelihood ratio means a greater probability that the variable is likely to have an influence greater than zero (zero representing the null hypothesis).

Relative risk is the probability of choosing one outcome category over the probability of choosing the baseline category. The risk ratios are more useful in terms of interpretation than the coefficients alone; both should be reported. For this, choose Odds Ratio from the Model Coefficient section.

Model Coefficients - Age

Age	Predictor	Estimate	SE	Z	p	Odds ratio
25-29 - <25	Intercept	0.264	0.163	1.62	0.106	1.302
	Education:					
	Low – High	0.444	0.174	2.55	0.011	1.559
	WantsMore:					
	Yes – No	−0.569	0.171	−3.34	< .001	0.566
	Usage:					
	Yes – No	0.411	0.176	2.34	0.019	1.508
30-39 - <25	Intercept	0.553	0.152	3.64	< .001	1.739
	Education:					
	Low – High	1.517	0.159	9.53	< .001	4.558
	WantsMore:					
	Yes – No	−1.293	0.159	−8.12	< .001	0.274
	Usage:					
	Yes – No	0.939	0.165	5.68	< .001	2.557
40-49 - <25	Intercept	−0.882	0.210	−4.20	< .001	0.414
	Education:					
	Low – High	2.364	0.214	11.03	< .001	10.636
	WantsMore:					
	Yes – No	−2.115	0.217	−9.76	< .001	0.121
	Usage:					
	Yes – No	1.184	0.215	5.50	< .001	3.269

There is more than convention in the reporting of both the coefficients ('Estimate') and the Odds ratios. The former shows the direction of effect, with a negative or positive number, while the latter allows us some idea of proportionality. Note that confidence intervals can be shown as well.

Based on the direction of the coefficients, interviewees in their twenties were most likely to want more children, least likely to be using contraceptives and relatively well-educated. Those in their forties were characterized by being the least well-educated and rather more likely to use contraceptives. Interviewees in their thirties were in an intermediate position.

You may also discuss the variables individually in terms of risk ratios. A sample interpretation is as follows: The relative risk ratio for having a low education level as opposed to a high education level is 10.6 times within the 40-49 age bracket than for women in the under 25 years old category. This means that the respondents who are in their forties are less well-educated than those respondents in their twenties.

You can also examine the variables graphically using the Estimated Marginal Means section.

Ordinal logistic regression

Madsen (1976) studied different types of housing in Copenhagen, covering areas selected to minimize variations in social status. The variables included the feeling of influence over how the accommodation was managed, the degree of contact with neighbors, the type of housing and satisfaction with the housing conditions. The file, Copenhagen.csv, contains 1,681 cases, and has already been converted into 'raw' data for you from the original summary data.

Satisfaction	Influence	Contact	Housing
low	low	low	tower blocks (coded 'tower')
medium	medium	high	apartments
high	high		atrium houses ('atrium')
			terraced houses ('terraced')

Open Copenhagen.csv; use Regression / Logistic Regression / Ordinal Outcomes. Satisfaction is to be our outcome ('Dependent') variable here, with the other variables as predictors ('Factors').

First, we need to think about releveling the dependent variable (outcome). *It is extremely important* in running ordinal regression properly. Press the arrow button at the top right of the dialog box to show the data file. Select the dependent variable column header, in this case Satisfaction; then either double-click the title header of the column or go to the Data tab and select Setup.

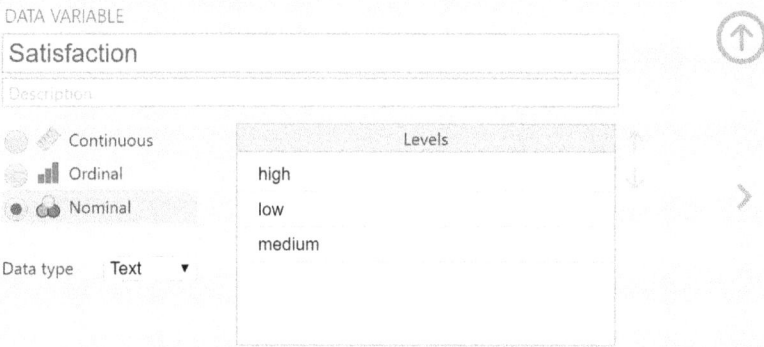

As becomes apparent, the program orders the levels in alphabetical order.

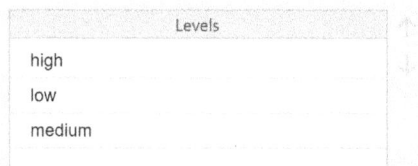

What you need to do (if you don't want useless results), is to order them in a hierarchical fashion. Select a level and move it using one of the small vertical arrows on the right.

Continue until you have a meaningful hierarchy of levels.

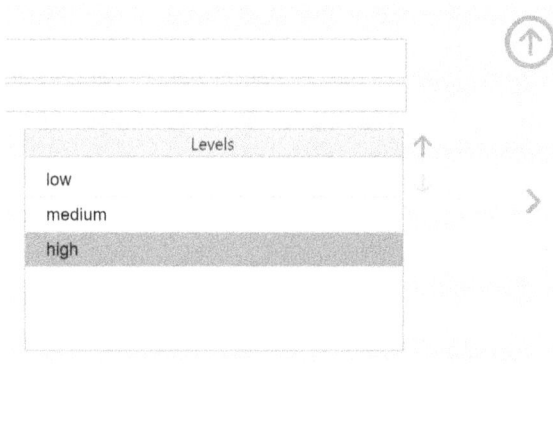

We now have a meaningful order of levels. You can then leave the dialog box by pressing the large vertical arrow at the top right.

Do note that the order – here, low-medium-high versus high-medium-low – does affect the results when ordinal regression is run. You will know when you've got this wrong: the Estimate statistics for an obvious effect will be negative when they ought to be positive or vice-versa. In this example, the relationship between high and low Influence in terms of Satisfaction should clearly be positive, not negative. If you do find an anomaly, this is a problem, as it also affects the odds ratios, so you need to reverse the order of the hierarchy of levels and let the program re-calculate.

This reminds me of a more general point. If you have made some changes to the settings, for example removing variables from a model, it might be worth re-opening the file and creating a new session of ordinal regression to ensure reliable results.

Now carry out the test by pressing the Regression button and selecting Ordinal Outcomes.

I have opened the Reference Levels section immediately: the setting of baselines makes a difference, as you will only see the variables with which they are compared. In this example, let us say that we already know about tower blocks and are more interested in the results from other forms of accommodation. We want to know which effects are positively linked with satisfaction and to what extent, so we choose baselines of low contact and a perception of low influence.

Model Coefficients - Satisfaction

Predictor	Estimate	SE	Z	p
Housing:				
apartments – tower	−0.572	0.1192	−4.80	< .001
atrium – tower	−0.366	0.1552	−2.36	0.018
terraced – tower	−1.091	0.1515	−7.20	< .001
Influence:				
high – low	1.289	0.1272	10.14	< .001
medium – low	0.566	0.1047	5.41	< .001
Contact:				
high – low	0.360	0.0955	3.77	< .001

If you followed the instructions to the letter, you should see these results. The program alerts you to which way you have ordered the dependent variable, lest you forget. We see the usual regression output coefficient table which includes the value of each coefficient (Estimate), standard error, z and p values. Note that the p value is a rather rough indicator in logistic regression.

Omnibus Likelihood Ratio Tests

Predictor	χ^2	df	p
Housing	55.9	3	< .001
Influence	108.2	2	< .001
Contact	14.3	1	< .001

From the Model Coefficient section, you could select the Likelihood ratio tests, which indicate that all three independent variables are significant in their relationship to Satisfaction. The table also demonstrates their relative influence.

You can also refer to the estimates' confidence intervals, which appear to be either positive or negative for each set of levels, staying on one side of zero. All of these results are likely to be statistically significant. The reverse cannot be assumed, however: if the confidence intervals for a variable straddle zero, *this does not necessarily* mean that the result is non-significant; you would need to investigate further. *

The coefficients from the model and their confidence intervals can be somewhat difficult to interpret because they are scaled in terms of log odds. As previously, it is more meaningful to look at the odds ratios.

Model Coefficients

Predictor	Odds ratio	95% Confidence Interval	
		Lower	Upper
Housing:			
apartments – tower	0.564	0.446	0.712
atrium – tower	0.693	0.511	0.940
terraced – tower	0.336	0.249	0.451
Influence:			
high – low	3.628	2.832	4.663
medium – low	1.762	1.436	2.164
Contact:			
high – low	1.434	1.189	1.730

The log odds in this (adapted) table show that a feeling of high influence is likely to raise the likelihood of feeling satisfied with accommodation by about three times. Other factors are also likely to cause satisfaction, but with less effect, and it should be remembered that these are factored in together as an overall model.

Within a model, some effects are interactions. You may want to use other tests to examine, for example, the relationship between Housing and Influence. Log-linear regression analysis and the contingency tables for the Chi square test of association may prove informative.

---

*It should also be noted that confidence intervals are currently under attack for not assigning probability to data. Morey et al (2016) say that these "generally lead to incoherent inferences" and recommend their abandonment for the purposes of inference.

Outputs / outcomes / dependent variables	Test
Two	Binomial logistic regression
Three or more	Multinomial logistic regression
Three or more, can be ranked (ordered)	Ordinal logistic regression

Chapter 15 - Partial and semi-partial ('part') correlations

Our study of correlations in Chapter 6 used **raw** correlations. These are also known as **gross**, **zero-order** or **unpartialed** correlations. These can be misleading, as any relationship between two variables can be affected by a third variable (or more).

Partial correlations

Partial correlation allows you to study the relationship between two variables while holding one or more other continuous variables (covariates) constant. Put rather better, the partial correlation is the association between two variables after eliminating the effect of a fraction of other variables (Kim 2015).

Relationships between phenomena are not always what they appear to be. As Mark Twain said when asked if he was worried about the levels of mortality in train crashes, you really don't want to go to bed, so many people die there. Let us consider some serious reasons for **spurious** correlations, the sort of problems that 'partialling out' or 'controlling for' variables might solve.

Unnecessary variables: One example was the mistaken idea that watching a lot of television had a direct impact upon academic performance. The more valid relationship was between television and IQ: youngsters with less intelligence were more likely to watch television for prolonged periods of time (and vice-versa). When IQ and other relevant factors were controlled for, the relationship between academic performance and heavy duty tv-watching disappeared.

Mediating (also known as **intervening**) **variables**: It is possible that one variable affects another, which then has an impact on a third. So just as it is possible to have unnecessary variables, it is also possible to omit necessary ones. For example, highly educated people may spend more than less educated people, but the important missing variable is likely to be income. More education is likely to mean a higher income which makes it easier to spend more.

Moderating variables: A relationship between two variables may not hold for all categories of another variable. An apparently straightforward correlation between a teaching method and student performance may, for example, vary with the gender of the students.

Multiple causation: Some phenomena have more than one cause. It is possible, for example, that certain types of road accidents only occur under certain road conditions and with a particular attitude to driving.

We therefore '**partial out**' or '**control for**' variables to deal with these problems. As usual, the context of the study should be your main guide. But in general, you may find the following strategies helpful.

If you think you may have an unnecessary variable, then you may see if a correlation of two other likely variables survive its influence as a covariate. In our televisual example, IQ and educational performance would be unaffected by the influence of duration of television-watching.

To look for mediating variables, you may look to see if a correlation becomes much smaller when the mediating variable is introduced as a covariate. The education-expenditure correlation should become smaller when income is taken into consideration. Do note that you can also encounter **suppressor variables**. In such a case, the correlation becomes much *bigger* when a covariate is introduced. This is a controversial topic, but it can mean that there is another factor which works in a peculiar direction. You may find, for example, a positive relationship between attraction and obese men; this might make more sense when body-mass-index is introduced (big men are favored, but this is moderated by their shape).

With regards to multiple causation, this is fairly obvious. You examine different correlations, and see if other variables make them bigger.

In all of these cases, knowing the subject area, or domain, is important. You might use the partial correlation approach as a way of deciding upon likely models to be examined by multiple regression.

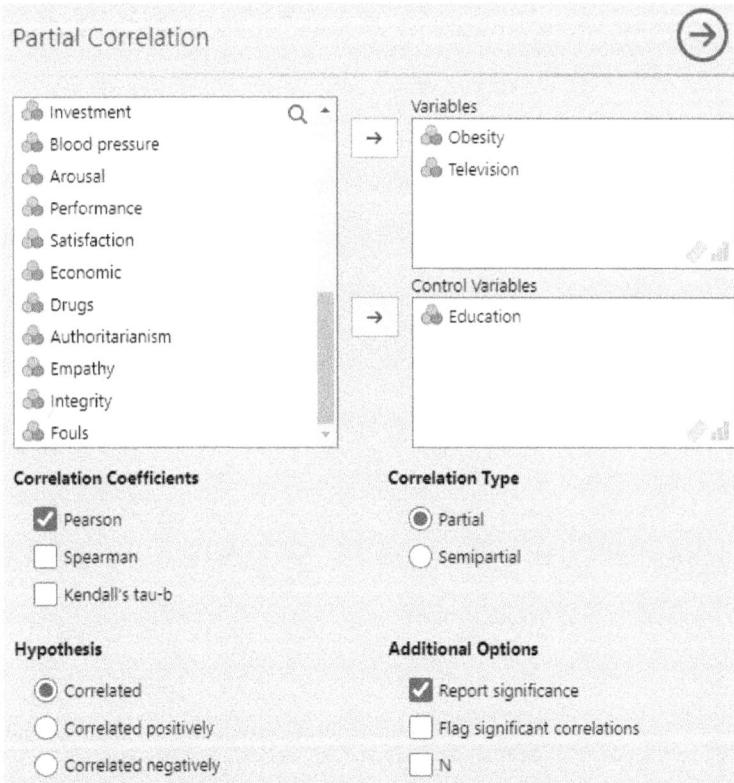

Open Correlations.csv in Jamovi, press on the Regression tab and choose Partial Correlation from the drop-down list. In the chapter on correlations, we found what appeared to be a moderate relationship between obesity and watching television, with Pearson's r at .6 and a p value of .051. What if we were to control for education?

Partial Correlation

		Obesity	Television
Obesity	Pearson's r	—	
	p-value	—	
Television	Pearson's r	0.869	—
	p-value	0.001	—

Note. controlling for 'Education'

With differing educational levels taken into account, television-watching and obesity are clearly related to each other.

Semi-partial correlations (also known as part correlations)

These provide the specific contributions of each covariate to the relationship. They are less helpful than partial correlations in explaining an interaction.

Semipartial Correlation

		Obesity	Television
Obesity	Pearson's r	—	0.795
	p-value	—	0.006
Television	Pearson's r	0.813	—
	p-value	0.004	—

Note. controlling for 'Education'
Note. variation from the control variables is only removed from the variables in the columns

There are no clear differences between the two variables in terms of their contributions to the relationship. Occasionally, one variable may appear to make a larger contribution than another, and some are tempted to argue that the semipartial explains the direction as well as strength of the variable's contribution. I would recommend finding more evidence on cause and effect rather than trying to squeeze too many drops out of the data.

Part 5

Bayesian statistics introduced

Chapter 16 – Theory: the minister, the prior and the post

Classical statistics – a brief preparatory overview

Without yet going into details, we will look at some of the differences between the two types of statistics. Up until now, we have been considering classical (sometimes known as 'frequentist') statistics, in which a sample is examined in the expectation that its characteristics will be replicated in the overall population.

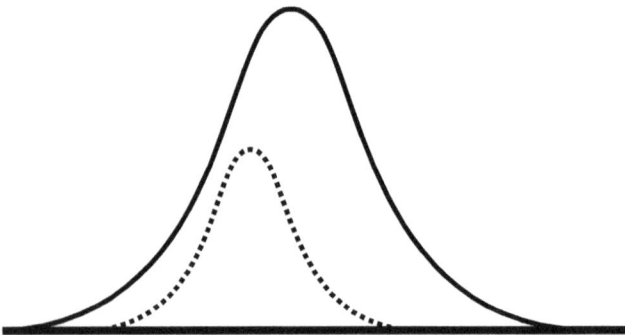

The basic assumption is that if one looks at sample after sample (hence the term 'frequentist'), each would appear quite similar.

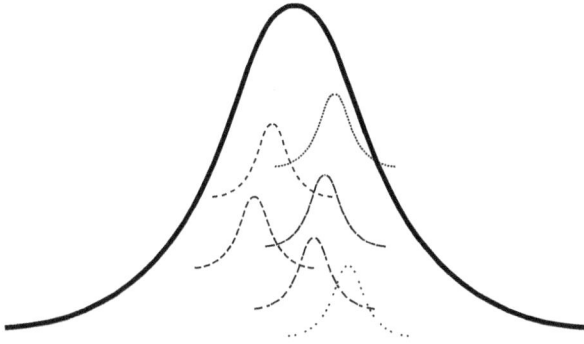

Related to this is the idea of a stopping rule. Not to go into too much fine detail, how would you feel if I said "well the experiment didn't prove my point, so I'll get more observations?" With your experience in classical statistics, you would probably think 'what an old fraud!'

Another point about classical statistics is that a large part of it is about rejecting the null hypothesis. While this is very much part of the Popperian view of science, that theories should be disprovable, it is nevertheless a touch convoluted and in statistics does not provide a precise negative: a test does not prove that your effect does not exist; it shows that a random or otherwise neutral situation cannot be disproven.

Yet another point is that 'significance', more respectably the rejection of the null hypothesis, seems pretty much an arbitrary affair: it is, or it isn't. Now when you think about it, this is a peculiar state of affairs. Let's take a *reduction ad absurdum* example. Suppose we have some reason to believe that over time, the absorption of literary fiction creates more considerate, less violent people. While the p value is a perfectly reasonable thing to quote, giving you some idea of the evidence against the null hypothesis, at some point the reader wants to know, "well is it significant or not? you know, at $p < .05$?" (or whatever is the critical value). Are my children going to have more empathy if they just read a bit more Dickens and Annie Proulx? Put at its worst (and if we ignore issues such as effect size and the pragmatism of RA Fisher), there seems to be an absolutism about classical statistics: it's significant or it's not.

Bayesian statistics as the antithesis of classical statistics

I have rather moved into devil's advocate role. Again without going into detail, let us see in what ways the Bayesian approach to statistics can be viewed as a radical alternative to the classical tradition.

Bayesian statistics is not based on the relationship between a sample and a population. The frequentist population is a hypothetical one: you have to guess that your population is uniform enough to allow predictive sampling.

Bayesian statistics is based on probability, dealing with the evidence to hand. The relative unimportance of sampling means that there is no 'stopping rule'. If the population is irrelevant, so is the size of the sample. Let us imagine for example that in classical statistics you need to have 2000 cases for the purposes of a survey; after you have the sample, you study the relevant effects. One of the advantages of the Bayesian method is that you do not need to have the whole sample before you start. If you have, say, 300 cases, and it becomes obvious that the effect under consideration clearly doesn't exist, then you can stop. If you find that there is a tendency towards its existence, then you can gather more evidence.

Evidence is gathered with a much more straightforward logic than in classical statistics. You have evidence in favor of the null hypothesis and you have evidence in favor of the alternative hypothesis, the effect.

As the evidence accumulates, Bayesian statistics does not give us the 'significant or not' dichotomy. It provides a graded view of the evidence: how much is there? You will find in the reporting tables to be introduced shortly that there is something of an equivalent to 'non-significant', but the evidence itself is measured.

Bayesian statistics introduced, via conditional probability

Bayesian statistics considers probability in terms of calculated likelihood, working out how likely it is that something will happen again. Basic examples of what I have in mind are working out how many balls of a different color are likely to be in a bag when one is removed; how likely it is that a person is going to be affected by an illness; the odds of a sports team winning when we know the strengths and weaknesses of the team and its opposition.

Well, that is conditional probability, the probability of an event after another event has taken place. For example, the removal of a red ball from a bag affects the probability of how many balls of different colors are left in the bag. This is also called inverse probability.

By itself, inverse probability has a place in predicting, for example, the likelihood of a particular disease affecting people. It can even work out the probability of one particular set of people getting the disease.

However, Bayesian statistics gives conditional probability a twist in direction. Instead of seeing an actual event and calculating its effect on future events, Bayes is interested in looking at a rather uncertain situation first and then examining the effect on probability when new, known, information is added to the mix. Initially, we take a guess (of which more later), then we update our knowledge with fresh information (the sample) and see how it compares with our prior knowledge.

A brief history

Before we get involved in the details and terminology, it is worth having a quick glimpse at the history of Bayesian statistics. That way, you can see its applications, often vital in modern history, how it works, and that very important point, that it does work.

Thomas Bayes, a Presbyterian minister in eighteenth century Kent, took an interest in inverse probability. The preoccupation which led to having a branch of statistics named after him is the idea that probability is to some extent based on a type of belief, rather than hard knowledge in the form of frequencies. We start with some prior knowledge which forms a basic but incomplete theory of events. Then we add evidence.

For some reason, Bayes discontinued his study of this area of mathematics. * It was only after his death in 1761 that his friend Richard Price edited his notes and had them published. However, the first major use of this theory was by the remarkable French scientist Pierre-Simon Laplace, who apparently independently of Bayes, applied inverse probability to interplanetary movement, court testimony and other fields. It may be more accurate to call Bayesian statistics Laplacean, but it was perhaps a mark of his breadth of achievements that the French Isaac Newton abandoned inverse probability in favor of other scientific interests.

Bayesian statistics fell into disuse among academic statisticians but continued to be applied to a range of problems. In the famous Dreyfus case, the principle of updating knowledge with fresh knowledge defeated the collation of coincidences as evidence:

> "Since it is absolutely impossible for us [the experts] to know the a priori probability, we cannot say: this coincidence proves that the ratio of the forgery's probability to the inverse probability is a real value. We can only say: following the observation of this coincidence, this ratio becomes X times greater than before the observation." (Darboux et al 1908)

It was also, among other applications, used in American actuarial mathematics, the statistics of smoking and lung cancer, and during the Second World War, Alan Turing's cracking of the Enigma code and the detection of the location of German U-boats. A particularly graphic example of Bayesian methodology was the use of a grid system in searching for a sunken American aeroplane carrying a nuclear bomb; as the search went on, such knowledge as was gathered, by eye-witness testimony and fragments, was built upon by subsequent searches of grid

---

*While his thought experiments refer to a table, his use of a billiards table is almost certainly mythical.

---

cells. Each patch of the sea was evaluated as somewhat more likely to be positive ('getting warmer' in the hide-and-seek sense) or negative ('getting colder').

Instead of the previous tale of a theory falling into disuse in academia, Bayesian ideas were fiercely opposed through much of the twentieth century, only to emerge as the new rage in the early twenty-first century. For an engaging narrative covering the fearsome rows – often fought by very awkward personalities – as well as the applied successes mentioned above, read McGrayne (2012).

How are Bayesian statistics used to test hypotheses?

To return to Bayesian statistics, we have our prior theory of the world, be it a guess, an expert opinion or no theory whatsoever, and we then combine this theory with the results of the study to gain a new, updated view of the world. To consider this in non-statistical terms, think for a moment about Columbus stumbling across America.

He found what we now call the West Indies – the name itself a result of his knowledge at the time – and believed he was in India. Despite evidence undermining this, he is said to have gone to his deathbed unconvinced of the alternative, that he had found a new continent. Evidence built up until Amerigo Vespucci made it almost incontrovertible. At this stage in our own knowledge, we don't even consider weighing the evidence for America's existence.

Evidence builds up on top of prior knowledge. What we know now is what we knew before with added new evidence. This form of learning is the key to the Bayesian viewpoint, the immutable relationship between rational belief and evidence as a key feature of science.

Before you feel all at sea, don't worry, we are not so far from shore. We come across the same practical problem with classical statistics: if you end up with 999/1000, do you test it? Nope. So similarly, if so much evidence has built up in 'the prior', to use a Bayesian term, then further calculation becomes unnecessary.

Now we can move to the formal statistics, but as usual without much in the way of formulae. The formal terms for what we have been discussing are the **prior distribution** (or simply the prior) and the **posterior distribution** (or posterior). The former is the statistical evidence we have at first. The latter is what we have after the fresh evidence has been calculated:

Prior distribution $\Longrightarrow$ Posterior distribution

At a sophisticated level, it is perfectly possible – especially with an automated system – for the combined knowledge above to become a new prior, meeting new information to create yet another posterior, and so on:

(Prior $\Longrightarrow$ Posterior) $\Longrightarrow$ new Prior

(new Prior $\Longrightarrow$ new Posterior) $\Longrightarrow$ another new Prior

Mathematically, there are computerized steps to be taken once the prior and sample are known. In particular, a **likelihood function** takes your sample data and analyzes the conditional probabilities. The likelihood function is very different from the prior. The Bayesian analysis then finds an intermediary distribution, which fits between these: that is the posterior distribution.

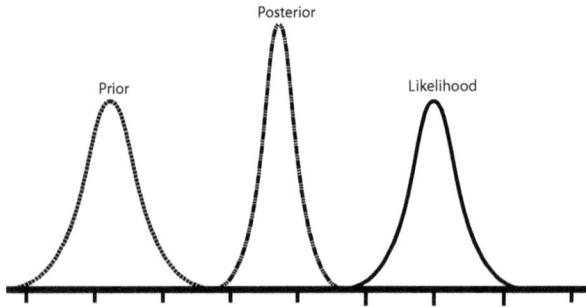

Generally, the preoccupation in Bayesian analysis is between the prior and the posterior, whether or not they are positioned differently.

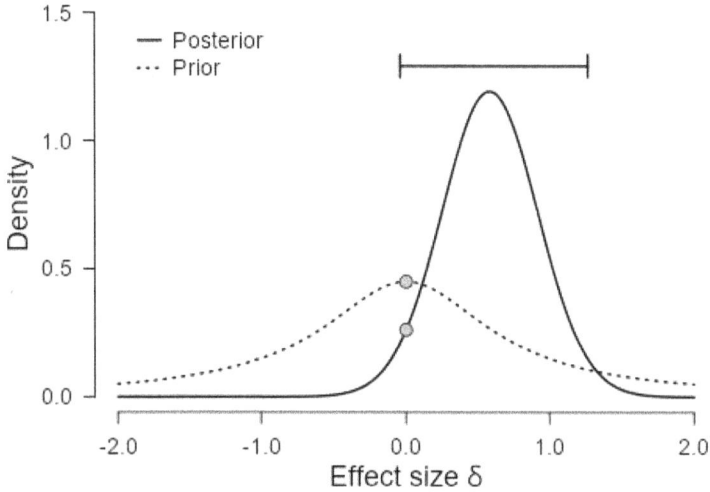

Here, in a chart from the Jamovi JSQ module (to be discussed in the next chapter), we see a clearly significant result. The posterior information is rather far away from the prior.

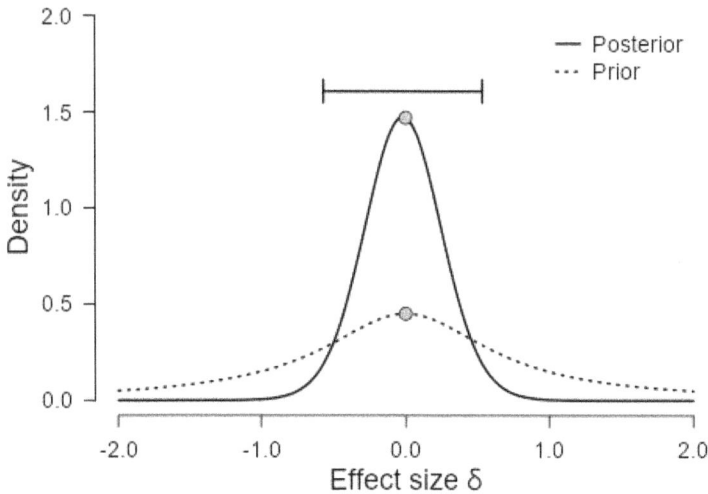

Here the two curves are positioned similarly, a clearly insignificant result.

Starting with different priors according to the researcher's current knowledge is almost definitively subjective. Some find this lack of objectivity disturbing; choice of other priors can lead to somewhat different results. On the other hand, it should be noted that the more observations you have, the less important the prior becomes. Also, to my limited mind, the reporting of the priors used should allow replication (just as I have recommended recording your options for other tests). But, see the next section..

And now, even better news!

In practice, people doing what you're doing in this book, testing hypotheses, don't need to devise priors. In general, you will assume a lack of knowledge of the world (with a suspicious similarity to the 'null hypothesis' of classical statistics). So we use **uninformative priors**, the default setting. This can be changed, but if in doubt, leave the default as it is.

For our purposes, the main differences between classical tests and Bayesian tests are the mathematics and their philosophy. On the subject of the mathematics, I will merely note that the recent appearance of computerized Bayesian analyses is no coincidence: Bayesian statistics are simple as fundamental logic, but demand a lot of processing.

Chapter 17 – Application: Jeffreys and Jamovi

Reporting Bayesian results

In classical statistics, we tend to use the p value and the related critical value (such as $p < .05$) to determine whether or not we can reject the null hypothesis, that variance is the result of noise and chance fluctuation. Bayesian statistics are reported more intuitively: evidence is reported *in support* of the alternative hypothesis (and the null hypothesis).

Bayes factors can be placed within reporting bandings, providing measured evidence. The following reporting suggestions are adapted from Jarosz and Wiley (2014):

Statistic		Quantification of evidence	
Bayes Factor (BF10)	BF reciprocal (BF01)	Jeffreys' interpretation (1961)	Raftery's interpretation (1995)
1 – 3	1 – 0.33	Anecdotal	Weak
3 – 10	0.33 – 0.1	Substantial (substantive, moderate)	Positive
10 – 20	0.1 – .05	Strong	Positive
20 – 30	.05 – .03	Strong	Strong
30 – 100	.03 – .01	Very Strong	Strong
100 – 150	.01 – .0067	Decisive	Strong
> 150	< .0067	Decisive	Very Strong

The Bayes factor is evidence in support of the alternative hypothesis. The reciprocal (less mathematically, its converse) is evidence in support of the null hypothesis. These are referred to as BF_{10} and BF_{01} respectively in Jamovi. (If you use the short-cuts on Jamovi's classical t tests, the 'Bayes factor' refers to the alternative hypothesis, BF_{10}.)

Large numbers in BF_{10} support the **alternative hypothesis**, that the effect being studied is significant; larger numbers in BF_{01} support the null hypothesis.

The lowest reporting level is a good example of how the grid differs from the classical interpretation of statistics in giving us gradations. We are given quite clear guidance: a Bayes factor of between 1 and 3 indicates evidence that is Anecdotal/Weak. In classical statistics, we would instead be scratching our heads with a p value of .052; maybe it is 'insignificant', 'a trend towards significance', or 'a tenuous result'.

In practice, results which narrowly attain the classical $p < .05$ critical value typically appear in the 'Substantial' Bayesian banding. This is not always the case, as the two sets of calculations do different things. Lee and Wagenmakers (2013) prefer 'Moderate' to 'Substantial', as the latter seems rather strong (Schonbrodt, 2015). It is of course possible that Jeffreys meant 'substantive', meaning of importance in the real world ('substantial' means of some considerable size). 'Substantive' might fit this category, but as the word is not commonly used and easily confused with 'Substantial', the word actually used in Jeffreys (1961), Lee and Wagenmakers' suggestion of 'Moderate' seems most sensible.

Considering the third category, Rafter considered that there was no evidence to justify Jeffreys' use of 'Strong' here. So he prefers to stay Positive! He selects the fourth category as the threshold for 'Strong'.

Not present is a reporting band for Bayes factors (BF_{10}) of less than 1 and reciprocals (BF_{01}) of greater than 1. Indeed, E-J Wagenmakers, the founder of the JASP statistical package, would prefer there to be a 'noise' category. In counterpoint, his colleague Richard Morey says "a number's a number; why categorize?" (I'll leave you to decide on an answer to that.) Results in this 'noise' area would be dismissed as clearly non-significant.

I still think that, regardless of which test is used, issues such as context and effect size should be considered. But one road I do not wish to go down is that of quantifying the relationship between the Bayes factors and the two hypotheses, as in 'this is so many times bigger than that'. To do this we would need to distinguish between the mathematical concepts of likelihood and probability, and the situation pertaining to the prior odds. You can see on the internet people digging around the fine distinctions of such statements. Apart from suggesting that, of the pundits, E-J Wagenmakers is probably the clearest, I would say that for the non-mathematically inclined, that is not a good place to go!

Installing the jsq module in Jamovi

To install Jamovi modules, press the Modules cross on the upper right.

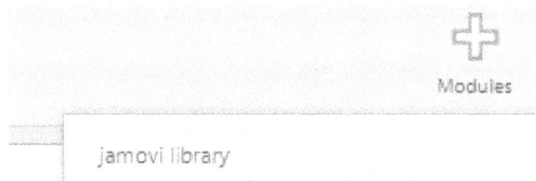

When you have chosen the 'jamovi library' drop-down item, you should be able to find the **jsq** module:

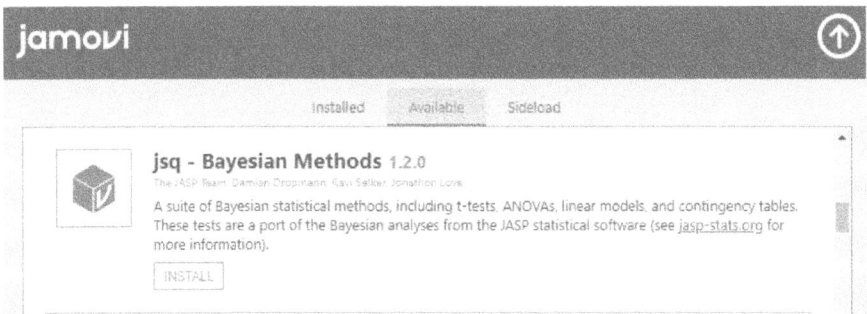

These menus contain Bayesian alternatives to the classical tests:

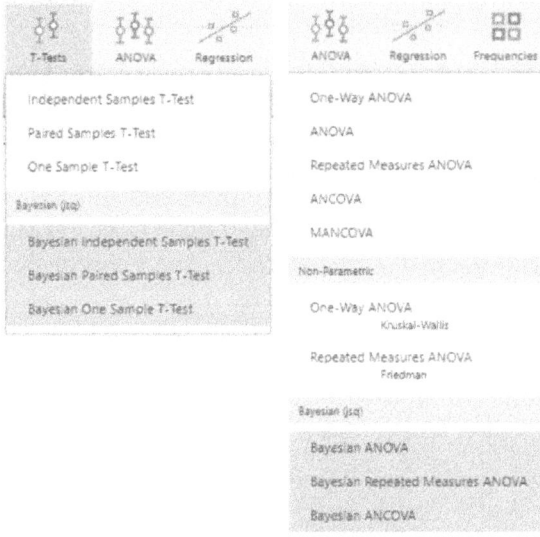

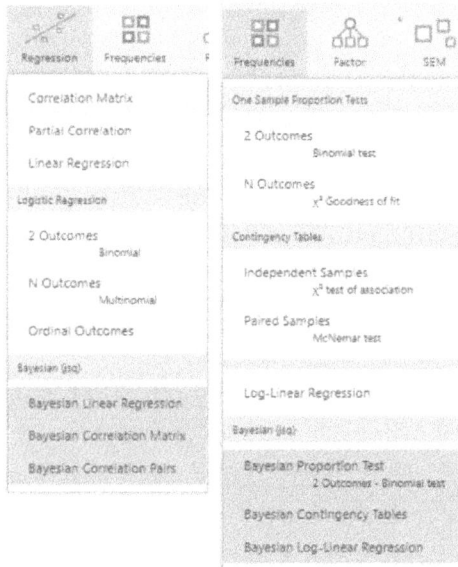

A practical example using the paired *t* test

Let's return to Differences.csv and look at the differences between the first three variables, use of alcohol affecting referees taking tests of judgement. On this page, I show the set-up and results of *classical* paired *t* tests, two-tailed, showing three rather different results.

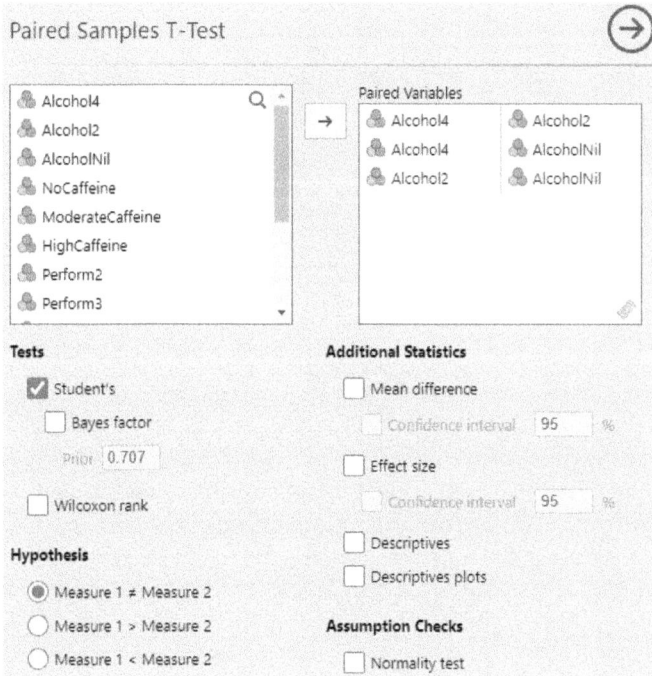

Paired Samples T-Test

			statistic	df	p
Alcohol4	Alcohol2	Student's t	−0.101	9.00	0.922
	AlcoholNil	Student's t	−3.065	9.00	0.013
Alcohol2		Student's t	−2.097	9.00	0.065

Bayesian Paired Samples T-Test ⊕→

Alcohol4	Q ^		Paired Variables	
Alcohol2	→		Alcohol4	Alcohol2
AlcoholNil			Alcohol4	AlcoholNil
NoCaffeine			Alcohol2	AlcoholNil
ModerateCaffeine				
HighCaffeine				
Perform2				
Investment2	▼			

Hypothesis

◉ Group 1 ≠ Group 2
◯ Group 1 > Group 2
◯ Group 1 < Group 2

Bayes factor

◉ BF_{10}
◯ BF_{01}

Additional statistics

☐ Descriptives

Plots

☐ Prior and posterior
☑ Additional info

☐ Bayes factor robustness check
☑ Additional info

☐ Sequential analysis
☐ Robustness check

☐ Descriptives plots
Credible interval 95 %

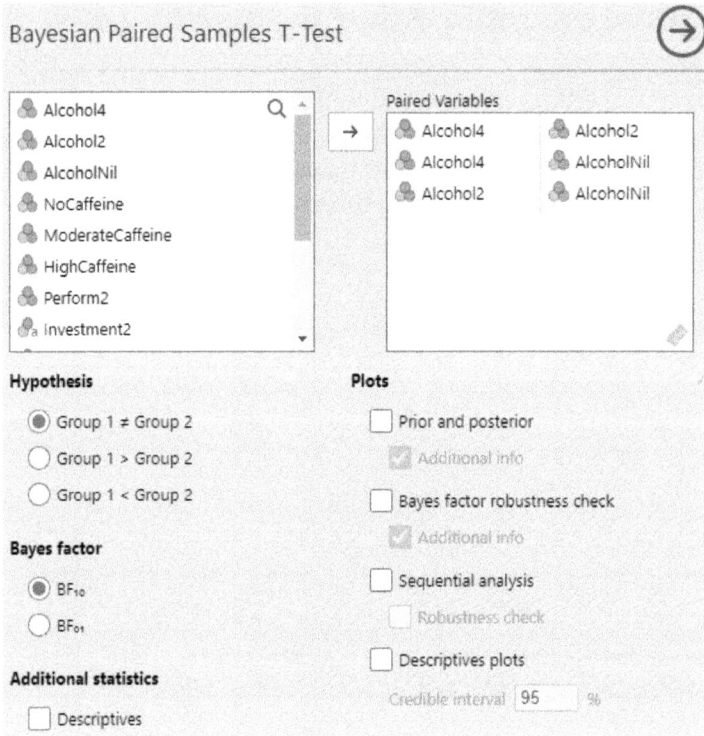

The set-up for the *Bayesian* equivalent of paired *t* tests is quite similar to that of the classical *t* tests. We will use the default Bayes factor setting of BF_{10}, testing the alternative hypothesis. (BF_{01} provides evidence in favor of the null hypothesis, as featured in the second column of the reporting table shown previously.)

Bayesian Paired Samples T-Test

			BF_{10}	error %
Alcohol4	-	Alcohol2	0.310	0.00707
Alcohol4	-	AlcoholNil	4.892	3.43e−4
Alcohol2	-	AlcoholNil	1.441	0.00343

Unsurprisingly given the *p* value, the Bayes factor for the first pairing is miniscule. The Alcohol4-AlcoholNil pairing has a Bayes factor of almost 5, moderate evidence in support of the effect. The Bayes factor for the final pairing is within the anecdotal/weak banding, in keeping with the rather tenuous classical *p* value of 0.065.

What if we had expected the direction of the effect, adopting a one-tailed hypothesis for the *classical* paired *t* tests? In this instance I chose the Measure 1 < Measure 2 option. (For good measure, I include effect sizes.) The status of the final pairing changes, reaching the $p < .05$ critical value:

Paired Samples T-Test

			statistic	df	p		Effect Size
Alcohol4	Alcohol2	Student's t	−0.101	9.00	0.461	Cohen's d	−0.0319
	AlcoholNil	Student's t	−3.065	9.00	0.007	Cohen's d	−0.9693
Alcohol2		Student's t	−2.097	9.00	0.033	Cohen's d	−0.6633

Note. H_a μ Measure 1 - Measure 2 < 0

The *Bayesian* paired *t* test Hypothesis section is set up as Group 1 < Group 2 (with additional plots and information):

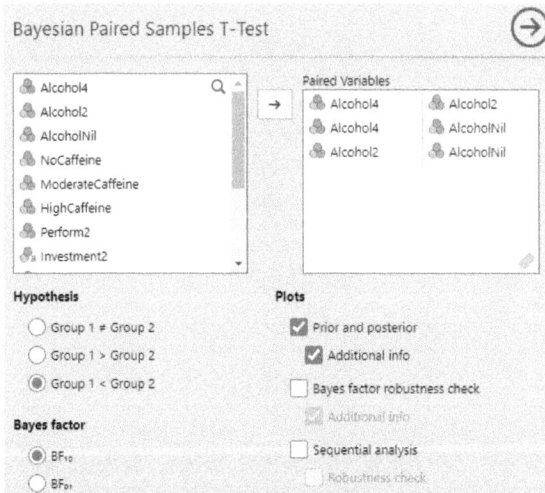

Bayesian Paired Samples T-Test

			BF$_{-0}$	error %
Alcohol4	-	Alcohol2	0.333	0.00163
Alcohol4	-	AlcoholNil	9.674	3.92e-5
Alcohol2	-	AlcoholNil	2.755	1.71e-4

The Bayesian analysis is less impressed with the final pairing, continuing to categorize the evidence as anecdotal/weak. As can be seen, Bayesian and classical tests do not always produce analogous results.

One interesting feature demonstrates the proportions of evidence for the alternative and null hypotheses as examined by Bayesian tests. If you choose the 'Prior and posterior' plot and also 'Additional info', you will see a rather nifty 'pizza' chart indicating the balance of evidence in favor of H1, the alternative hypothesis, and in favor of H0, the null hypothesis, or 'noise'. BF$_{10}$ is the Bayes factor in favor of the alternative hypothesis; BF$_{01}$ is the relevant statistic in favor of the null. Here I refer to the first pairing, using a two-tailed test.

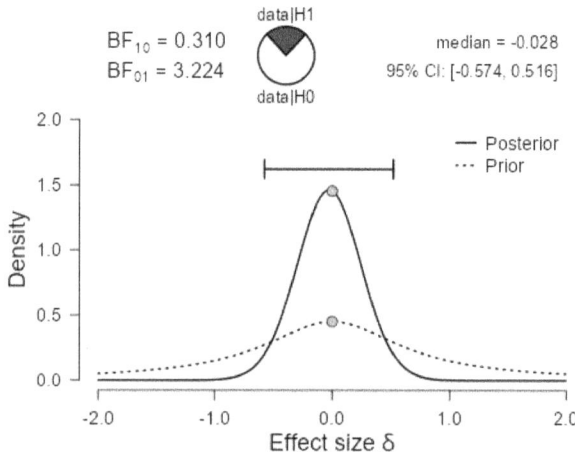

The dark shade of the pizza indicates rather scanty evidence for the alternative hypothesis (H1) with a lot more for the null hypothesis (H0).

Two important points must be made:

Bayesian tests in Jamovi have similar *assumptions* about data as their *parametric* counterparts.

The Prior section default is the uninformative prior setting of 0.707. I recommend that you *use the default* (unless or until you have a more sophisticated understanding of Bayesian analysis).

At the time of writing, Jamovi's jsq module has the range of Bayesian tests as shown in the menu images. It should be noted that while these tests have familiar names, such as ANOVA, the names have been chosen so that their overall role is easily recognized. They are not completely analogous to the classical tests, being different instruments using different calculations.

Which tests to use, classical or Bayesian?

Often, the tests come out with similar results for the same set of data.

Many proponents of Bayesian statistics say that you do not have to jump through that rather convoluted hoop of classical statistics, the rejection of the null hypothesis (the negation of a negative). Also, Bayesian tests do not consider the sample against a largely unobserved population (Wagenmakers 2007), but considers the sample in itself, in combination with the known (actually, usually uninformed) current knowledge. This being the case, one should be able to refer to evidence in favor of the alternative hypothesis, the theory of interest to the researcher, or perhaps more profitably, look at different alternative hypotheses, such as X does not equal Y, or more specifically X is bigger than Y (or vice-versa).

Conversely, this does suggest one really good reason for using classical statistics. If you wish to determine whether or not a hypothesis is true, then the classical/frequentist model makes more sense. According to Popper (1968), in science one can only falsify a theory. But we can ascertain if there is support for the null hypothesis. The p value represents merely the likelihood that the null hypothesis can be upheld, the major focus in classical statistics.

On the other hand, perhaps you have already rejected the null hypothesis and want to determine a model's likely **credibility** given the data. Then the Bayesian approach gives clear(ish) reporting categories (see the reporting table). So, for example, if you have a p value of .017, but are not sure of its level of credibility, you may seek a Bayes factor.

A common practical situation in classical statistical testing is the quandary of the uncertainty of a p value close to .05. Do note that its proponent, R. A. Fisher, did not consider this critical value to be an absolute, but a useful place to consider a hypothesis. Let us assume that you have not been dredging (lots of meaningless testing) but have a result which seems a little uncertain: you might wish to complement the classical test with a Bayesian test.

A more philosophically reasonable use of Bayesian techniques than the triangulation just suggested is the comparison of models. A variety of samples may be subjected to examination and it seems reasonable to compare their credibility by reporting their respective Bayes factors.

As this rather contrapuntal discussion suggests, you can to a large extent select the method you prefer. While they are different and provide somewhat different insights, there is no hard and fast rule that says 'use classical for X and Bayesian for Y'. While the null hypothesis may be considered to have its spiritual home in classical statistics, it can be tested using Bayesian methods.

So you can use the methods with which you are most comfortable, or which make the most sense to you. However, you will find different tools available. For example, the classical tests provide measures of effect size, how much of the variance can be attributed to the effect. Also, if you are dealing with non-continuous data such as ordinal data, I would stick to good old non-parametric tests; you might want to investigate resampling techniques such as permutation tests and bootstrapping, but that takes us well beyond the remit of this book.

A final consideration is what could be called Stats Wars. The internet is awash with articles telling you how Bayes is the way forward and how 'frequentists' (users of classical statistics) are pseudo-objective, have made lots of terrible errors and are just *so* twentieth century. * There are so many of these that I see no point in citing them, but would suggest that their quantity is probably related to the convergence of the wide use of the internet and the speed of twenty-first century computers. For a forceful but well-considered article in favor of Bayesianism and against commonly held frequentist errors, try Wagenmakers (2007).

There are, however, plenty of statisticians who remain frequentists. See, for example, Dennis (1996) and articles by Mayo (2012) and Steinhardt (2014), who suggests that the main reason why so many frequentists have produced 'bad science' is because most science *per se* has been conducted by frequentists.

A short article on the internet by Gelman (2012) seeks to decouple myths dividing frequentists and Bayesians. Among other things, he indicates that Bayesian statistics and testing the null hypothesis are perfectly compatible.

Critical thinking, as usual, is more to the point. So, you can use either set of methods or both. But I leave the final words to Frankie Howerd – a viewing of some of the *Up Pompeii!* series should convert you to innuendo and the art of the comic aside – "Oh suit yourself!"

---

*The more ebullient proponents of Bayesian statistics consider the 'null or not' way of thinking as being unworldly: "your test suggests that the world won't exist in the morning – I would bet that it will".

Part 6

Visual exploration:
The time until events
Cluster analysis

Chapter 18 – Survival analysis: the Kaplan-Meier curve

Introduction

This is the study of observations as a series of events. Unlike the content of the preceding chapters, observations are not measured according to their magnitude, nor are they qualitatively different from each other. Here, we are interested in a single category of data occurring over time, although we do have the possibility of comparing different groups' experiences. Our practical concerns are when things occur and their frequency at different times, while making allowance for missing observations.

In sociology and economics, you may find it referred to as 'event history analysis', 'event structure analysis', 'duration analysis' and 'duration modeling'. I use the term 'survival analysis' here because it is the most commonly used term for the techniques in statistical literature; you would refer to this if you were to delve into technicalities.

The popularity of the term 'survival analysis' derives from its common usage in medicine, where the death of patients appears to be a popular cause for concern. However, this terminology fails to reflect the wide range of possible applications of this set of techniques, which is reflected in its history.

One of the oldest areas of statistics, it started with seventeenth century studies of risk and patterns of longevity and mortality. Insurance and annuities came into being in an organized way, using life tables as the foundation for actuarial work.

This sort of study was later used in engineering to study how long it took for weapons to fail. Here, it became known as reliability analysis or reliability theory. Other names appear in different disciplines. This diversity of nomenclature may account for the absence of this type of analysis from general statistical introductions by other authors: superficially, it appears to be a specialist technique.

In fact, it has a broad usage. Events can be positive as well as negative. In sport, we could be interested in the length of time before rugby players become match fit, the event probably being the announcement of the decision. Or they could be transfers to or from a team, injuries, career milestones, successful completion of courses and dismissals.

These techniques are less precise predictors than the linear measurement–oriented techniques found in much of this book, but they are much more versatile. Linearity is not a concern here. Also, the techniques are incredibly adaptable, usable in tackling a wide range of problems in real life. You could study relegations or promotions of teams, violent incidents and perhaps the successful fruition of business or management operations. Perrigot *et al* (2004) examine the history of failures within business franchises in order to study what works organizationally and what does not.

Different categories can be contrasted. This could be the implementation of a new scheme in more than one type of setting, or against a 'control' setting where no such scheme is in place. Although care should be taken when comparing settings. Kahneman (2011) illustrates this in epidemiological studies involving urban and rural settings. Large apparent differences can appear because of small data sets, with rural fluctuations resembling nothing so much as a few throws of the dice.

Survival analysis is concerned with how long it takes for an outcome, or **event**, to take place. The focus is on the interval between a given starting point and a specified event. In sport, this can be the time it takes for adolescents to achieve particular standards in table tennis or judo. Negatives could include injuries, or defeats after (hopefully) a run of victories. One recent study examined influences on the length of time before donors defected from athletics programs (Wanless *et al* 2019).

The interval between the starting point and event has a variety of names (because we like to be clear in statistics!). It is referred to as the **survival time**, the **observation period** or the **follow-up period**.

The survival time is the area of examination. Do events tend to accelerate in frequency after a particular period of time? Are there phases in which events tend to cluster? What proportion of cases is affected in a particular phase?

We may also wish to contrast different samples. Does a group of footballers with participation in management have a different accident rate from a control group under a traditional management régime? Will the incidence of interpersonal conflicts differ according to the gender mix of the management team?

There are other ways of measuring event rates, for example moving averages or using the predictive power of linear regression. Survival analysis, however, has a range of advantages. It is largely descriptive, providing very informative analysis of the process being studied. Non--linear patterns are not a problem; mortality studies, for example, have inevitable shifts from the average at the beginning and end of the follow-up period. Crucially, however, survival analysis also accounts for missing information, known as **censored events**.

Some information cannot be observed; this is **censored** information. There are two types of censored information in the technical terms of survival analysis. One type is **left-censored information**, where a study needs to take account of information that is missing before the study takes place, whereas **right-censored** information is where information gets omitted during or after the study. Essentially 'left' means 'before' and 'right' means 'later':

$\Leftarrow$ left: before study | $\Rightarrow$ right: during and after study

Now for the sake of this chapter, we will ignore left-censored data. It is a fairly rare occurrence in studies and is not appropriate for an introduction to the topic.

Right-censored data, hence to be referred to as just 'censored data', appear in two main forms. One type is disappearance during a study: participants may withdraw from the study, move away, become incapacitated for the purpose of the study, or researchers may lose contact with them (otherwise known as losing the paperwork). The other type is when the event in question does not occur during the follow-up period. This '**loss to follow-up**', during or after the end of the study, means that we do not know whether or not the event happened.

The fact, for example, that a resignation or a promotion at work is not observed during the study does not mean that such an event has not happened. We just don't know. Even when we are certain that an event will take place (we are all dead in the long run, said Keynes), we cannot know *when*.

To merely omit censored data, or to consider this to be of the same duration as the last recorded event, is to underestimate the effect under investigation. To focus on events in isolation would be to miss the considerable richness of data provided by looking at the time preceding them.

To summarize, survival analysis, the study of the time until events, measures the occurrence of events in terms of the duration of time in which they take place. It also takes censored data into account in its calculations. Information which is missing either during or after the monitoring period, right censored data, is easily accounted for. You should avoid introducing 'left censored data', where complications precede the survival time; there are procedures for dealing with some of this, but this cannot be covered in an introductory book.

Statistical assumptions

Censoring should be random; do not exclude cases from the survival period because they appear to be a particularly high or low risk. The event must be categorical and dichotomous. Cases are of the same type and are either dead or alive, convicted or not, promoted or not, striking or not, and so on.

Methodologically speaking, time is the predictor variable and the criterion variable is the status (event occurred or not). Do not let that bother you, as the process itself is quite straightforward.

For the Kaplan-Meier function, otherwise known as the Kaplan-Meier estimator, the main method used in this chapter, continuous data are needed for the follow-up time, but as it is a non-parametric method, this is not a particularly exacting assumption. Although typically measured in hours or days, the stretch towards weeks or even months is possible as long as there is a fairly steady flow of data. Arbitrary intervals, such as years, are better handled by follow-up life tables (not covered in this book).

The Kaplan-Meier survival function

The beauty of the Kaplan-Meier is the intuitiveness of the survival plot: what you see is what you get. The Kaplan-Meier curve's steepness shows the likelihood of an event, whether or not it accelerates or decelerates over time, and even the differences between groups of cases under different conditions.

The minimum requirements are two variables. One is the follow-up time (or 'survival time'), measured in hours, days, weeks, or months from the start of the measurement. The other variable is dichotomous, either the event or the loss to follow-up of a case (censorship).

A small single cohort sample

Let's say that we are interested in hamstring injuries. We use the Injuries.csv file.

	days	recoveries
1	40	Recovery
2	60	Recovery
3	62	Recovery
4	80	Recovery
5	85	Recovery
6	108	Recovery
7	108	Recovery
8	115	Censored
9	140	Recovery
10	160	Censored
11	180	Censored
12	180	Censored
13		
14		

Typically, when survival analysis is applied to sports injuries, the 'event' is when the unlucky sportsperson or athlete suffers an injury, thus considering the length of time before an injury takes place. In this case, to demonstrate the positive usage of this type of analysis, let us consider recovery times after the diagnosis of a grade 2 or grade 3 hamstring injury (a tear as opposed to a strain).

The (fictional) study lasts for 180 days. Each cited event is a recovery from the injury. Censored data represents people disappearing from view, perhaps moving to another club or leaving the sport. The small data set is used for demonstration purposes, as is the overly short time period.

Let us consider the data. Two recoveries occur on day 108. On day 115, a participant is lost to follow-up for some reason; perhaps the prisoner was deemed no longer ill. There is another disappearance on day 160. Two more sportspeople have not recovered by the end of the study; they may or may not recover in the future, but we don't know and thus they too are lost to follow-up.

Our main interest is the pattern of events. The Kaplan-Meier curve, however, also takes into account the censored data in its calculations.

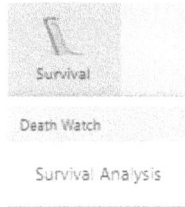

Open up *Modules* at the top right of the program, and from the jamovi library install Death Watch (one of the developers thinks it's funny). Press the Survival tab and then click Survival Analysis.

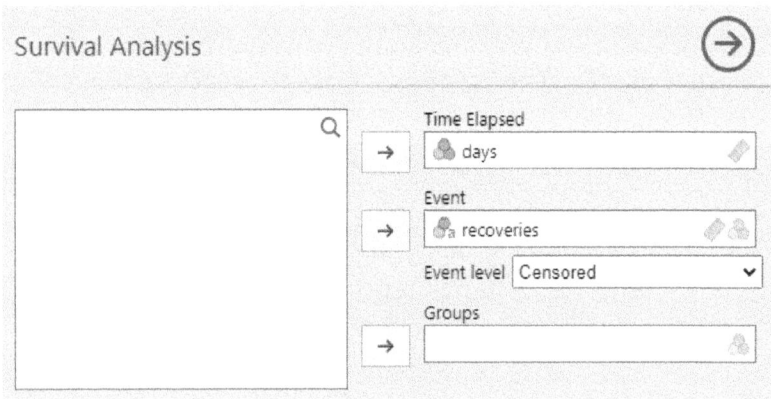

Put the time and event variables in the relevant slots. Unless the Event level begins with an A or B or something just before Censored, you'll get 'Censored' as the current 'Event level'. **You don't want that,** so toggle to the correct level:

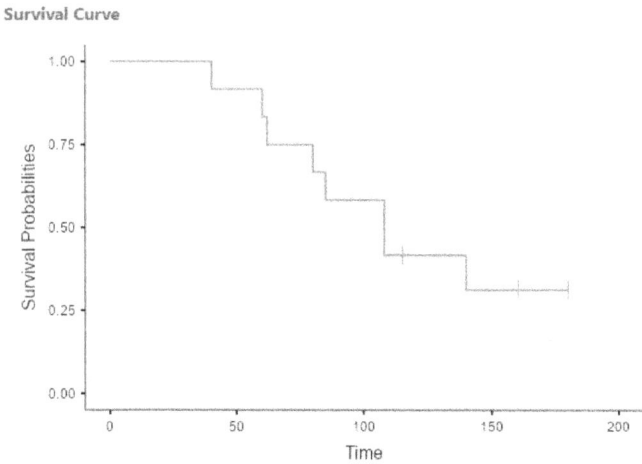

Events Summary

N	Censored	Observed Events	Expected Events
12	4	8	

Median Estimates

Median	Lower	Upper
108	80.0	NaN

Survival Curve

This is the Kaplan-Meier curve. Well, it gets to be a curve if you have a lot more events. Each step down reflects an event, with the steeper dip representing two recoveries on day 108. The notches are the censored events; if you don't want to see these, you can deselect the 'Censored events' checkbox.

The plot represents the likelihood of an event happening. The x axis represents the passage of Time. The y axis (Survival Probabilities) shows that at the beginning we have 100% of our cases injured. At day 40, after the first event, it looks like about 90% of the participants are still injured.

The median in the table above the chart is the 'estimated median survival time'. This should be quoted in any report. It is the length of time half of the cohort survive (or in this case, stay convalescent). You could see this on the chart by selecting 'Median(s)' from the Plots section, reading across the middle of the y axis to the curve and then looking downwards. Essentially, half of the cohort had recovered by day 108. Points 0.75 and 0.25 represent the first and third quartiles respectively. The median is preferred to the mean (which is given in SPSS), because highly skewed data is common in survival analysis.

A sample with two groups

We are interested in whether or not a course of boxing will help to prevent young offenders from committing further crimes. There are 30 young people in our study, in two groups. Those in the experimental group undertook the boxing course, while the control group underwent the usual youth justice procedures. The follow-up period is 90 days from the end of the course, with the event being reconviction. Loss to follow-up could be withdrawal from the course or a disappearance, but also those who survive the period without committing fresh crimes (and may or may not re-offend in the future).

	days	offending	cohort
8	84	Reoffend	Experiment
9	88	Reoffend	Experiment
10	90	Censored	Experiment
11	90	Censored	Experiment
12	90	Censored	Experiment
13	90	Censored	Experiment
14	90	Censored	Experiment
15	90	Censored	Experiment
16	16	Reoffend	Control
17	17	Reoffend	Control
18	18	Reoffend	Control
19	20	Reoffend	Control
20	22	Reoffend	Control
21	25	Reoffend	Control
22	25	Reoffend	Control
23	30	Censored	Control
24	53	Reoffend	Control
25	71	Reoffend	Control
26	84	Reoffend	Control
27	86	Reoffend	Control

This is a snapshot from the Rehab.csv file.

Remember to move away from 'Censored' to an appropriate Event level.

Let's examine the whole sample first, in order to show a little more of the Kaplan-Meier function.

Events Summary

N	Censored	Observed Events	Expected Events
30	11	19	.

Median Estimates

Median	Lower	Upper
83.0	54.0	NaN

Survival Curve

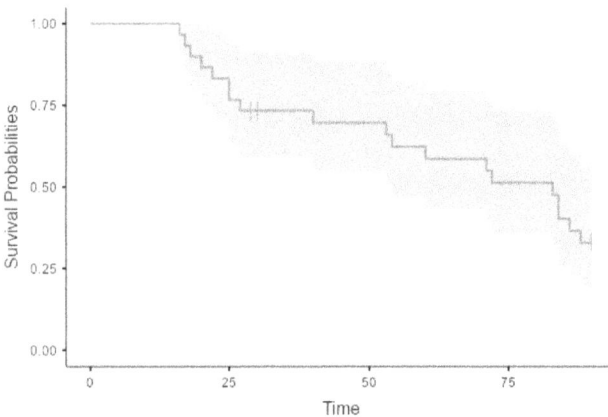

Here we see the confidence intervals, showing a range of typical values spread around the point estimate given by the median. Given the relatively high number of censored observations against a reasonably small data set, it is not surprising that the lines are quite wide apart.

There are few notches representing censored events. One such notch is right at the end of the curve; this suggests some loss to follow-up at the end of the study, that some people were not convicted by the end of the study. There is more than one way of interpreting this. It could well be that the study, at 90 days, is too short, especially as the estimated median survival time is 83 days. It could be that overall there is a low reconviction rate. Or perhaps the groups may show us something.

Events Summary				
	N	Censored	Observed Events	Expected Events
Control	15	4	11	7.37
Experiment	15	7	8	11.63

Median Estimates			
	Median	Lower	Upper
Control	53.0	22.0	NaN
Experiment	88.0	72.0	NaN

The summary table's usefulness now becomes clear. We can see that the experimental group has substantively more censored events than the control group. A higher than expected number of events (11 rather than 7.37) occurred in the control group; in this example, that means more re-offending. This is also reflected in the estimated median survival time. The experimental group 'survives' for a lot longer, half of them not re-offending close to the end of the study. Half of the control group have re-offended by day 53.

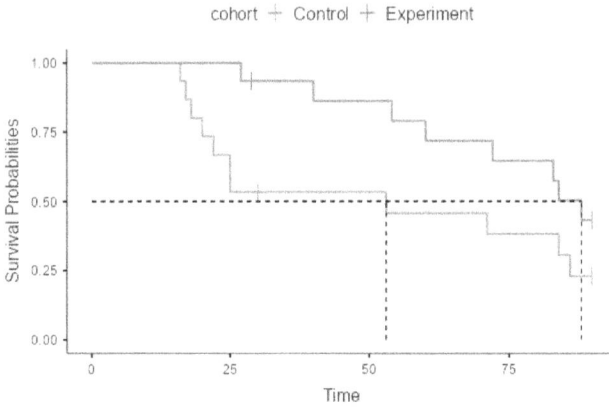

The curve shows an awful lot of offending in the lower grouping (delineated as Control; this is not clear in glorious black and white). The controls are out of control within the first month of the study. The dotted lines demonstrate the very different medians.

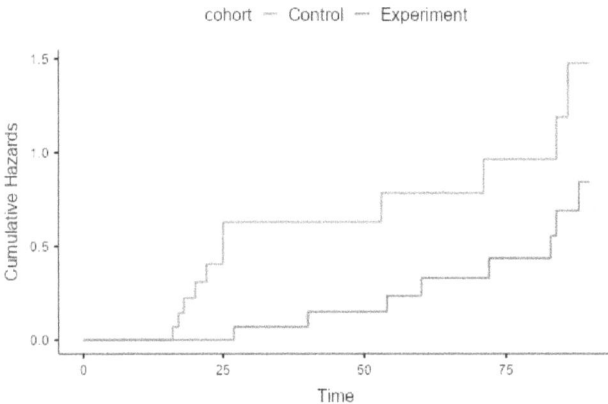

The cumulative hazard plot denotes the relative acceleration of risk. The longer the follow-up time, the greater the risk. As well as the steep trend of re-offending by members of the control group in the first month, there is a strong indication of vulnerability in both groups toward the end of the study.

Group comparison tests

Analysis of the differences

Test	χ^2	df	p
Log-rank	3.00	1	0.083
Gehan	4.49	1	0.034
Tarone-Ware	4.63	1	0.031
Peto-Peto	3.80	1	0.051

Now, here's the good news: you only need to apply one test for a pairing. And the bad news: you've got to decide which test to apply. And the goodish news: there is a reasonably clear set of rules for selection. The log-rank test gives equal weight to all time points; it is more sensitive when the two groups show consistently similar patterns. The Gehan-Breslow test is sensitive to groups in the early stages, but it can trend towards Type 2 errors (wrongly judging results not to be significant). The Tarone-Ware test is also heavily weighted towards the early stages, but is preferred where lines on the Kaplan-Meier plot intersect or move away from each other. If the hazard ratios of parallel groups are not proportional, then the Peto-Peto (also known as the Peto-Prentice) is considered more robust.

When we look at the survival plots and consider the descriptions above, the Tarone-Ware seems to me to be the most appropriate test for this data set. If we look just out of interest at all of the tests, the difference appears somewhat less likely to be significant under the log-rank test (.083), probably reflecting the differences between the two groups' patterns. Such differences hint at the dangers of dredging. For a very detailed recent analysis of the choice of tests in survival analysis, see the internet article by Karadeniz and Ercan (2017).

Jamovi can handle more than two groups. For each pairing, you would need to use the test which most reflects the relationship between their respective curves. Also, unless they are planned comparisons, you should apply a correction to avoid Type 1 error (considering results to be significant when they are not). This can be calculated relatively simply.

Let's say we had these p values for three pairings:
AB: 0.003, AC 0.03 and BC 0.02

The simplest method to calculate is the Bonferroni adjustment. You count the number of pairings, three in this example, and simply multiply the p values by the number of pairings:

AB x 3 = 0.009, AC x 3 = 0.09 and BC x = 0.06

In this case, assuming a critical value of $p < .05$, only the result of AB would justify rejecting the null hypothesis. However, the Bonferroni is considered to be overly harsh.

A more moderate adjustment, slightly more complicated but not mind-boggling, is the Holm-Bonferroni. Here you take the smallest of the values first and multiply that by the number of pairings. The next smallest value is multiplied by the number of pairings minus 1. The next is multiplied by the number of pairings minus 2, and so on.

So we would start with AB 0.003, BC 0.02 and AC 0.03

AB 0.003 x 3 = 0.009; BC 0.02 x 2 = 0.04; AC 0.03 x 1 = 0.03

Sorry about the mathematics.

Thinking point

As mentioned earlier, events can be positive or negative, and not necessarily life-threatening. Another point is that these methods stimulate research. As you pore over the tire tracks of time, new questions arise. This approach asks "a series of questions about the causal connections among actions... It relentlessly probes the analyst's construction, comprehension and interpretation of the event" (Griffin 2007).

This table only refers to Kaplan-Meier techniques.

N.b. *Non–parametric tests can be used with 'parametric' data.*

Test	Data	Purpose
Kaplan-Meier	Non-parametric Large and small sample Continuous time data	Tracking events over time
Log Rank Gehan-Breslow Tarone-Ware Peto-Peto	Non-parametric (for differences in usage, see earlier notes in this chapter)	Significance of Kaplan-Meier group differences

Chapter 19 – Cluster analysis

Introduction

Earlier, we looked at principal component analysis and factor analysis. In both of these, data reduction is achieved by a focus on the *columns* of our data sets, which represent the variables.

Cluster analysis also pertains to data reduction, but its focus is upon the *rows* of the matrix, which represent cases. Such cases can be observations of phenomena or individual survey participants. The data is reduced to groups, known as clusters. Within sales figures, we may discover that certain groups of people may tend towards different spending patterns. Attitudes to social issues could be different between, say, people of different ages or from different parts of the country. It is not unknown for groups to become newly constructed from surveys, as different types of consumer, elector or social type. So, in an inversion of factor analysis, we look not at core constructs but at the behaviour of clusters of individuals or groups of data (Everitt *et al*, 2011).

Cluster analysis is best suited to empirical work. Generally speaking, we do not have a governing theory, but think about what the clusters mean when we see them. We can follow up initial exploration with different ways of viewing the clusters, but cluster analysis is really a descriptive method, not an inferential one. Statistically speaking, it is not proof of anything! For an objective assessment of differences between groups, something like ANOVA or logistic regression would make sense.

While statistically of little use, cluster analysis is an intensely mathematical tool. It is what is called an 'unsupervised' classification method: this means that you let the machine get on with finding any hidden structure. By contrast, a 'supervised' method is basically where you have created a 'training' model against which to try out fresh data.

Central to cluster analysis is the concept of similarity. Cluster analysis works by calculating how elements within a cluster are similar to each other. Each cluster should be dissimilar to the others. Unlike PCA, we don't have stopping rules such Kaiser's criterion or the scree test. One recommended method is to wait until a 'jump' in results becomes apparent, then returning to the last preceding model. Another is, as usual, considering the rationale of the data being used.

There are two widely used sets of methods. These are 'hierarchical' and 'non-hierarchical' cluster analysis. The former constructs clusters in a tree-like fashion, while the latter merges or splits them.

Hierarchical cluster analysis is probably better when you have a theoretical idea of what clusters may represent, and is typically easier to interpret. The dendrogram is particularly intuitive, although a heat map and plotted variables are also amongst the tools.

The most commonly used non-hierarchical algorithms are referred to as k-means clustering. These are said to be more reliable than hierarchical methods, with more stable clusters, at least when applied to large numbers of cases, and are recommended for when there is little rationale within the data.

I personally would use hierarchical clustering for the initial examination of some small samples. Then it may be worthwhile to apply k-means clustering to a large sample to see if the insights generalize. The module's plotting tools may provide further insights.

A large body of literature has grown out of cluster analysis, while this is a short book. For a more detailed consideration of the methodology, try the chapter by Hair and Black (2000).

Data preparation

Outliers should be eliminated. These may cause 'noise', create clusters amongst themselves, or affect the formation of other clusters.

Multicollinearity also needs checking. Over-related items are, unsurprisingly, likely to cause clustering of an irrelevant nature. Whereas principal components analysis requires similar variables, cluster analysis needs similar cases.

Standardization can be used to remove response effects and differences in range between variables, and is often used for questionnaire data. The decision to use this option is a matter of judgement. If different measurements are non-meaningful in terms of your study – such as 5 point scales and 10 point scales, or kilometers and pounds – then you would use this. If all the measures are of the same type and the differences are meaningful, then don't standardize the data. If in doubt, as I am with the dataset used in this chapter, then use standardization. Standardization is a default setting in the tools used here.

Sample sizes have no strict rules. However, small samples of say 25 to 30 are considered best in hierarchical cluster analysis. *k*-means clustering is often used with samples of 50 or more, and also 'big data'.

Hierarchical cluster analysis

While it is probably easiest to interpret a group of 25 to 30 cases, the sample should be representative, so you will want similar behaviour from other batches. As noted above, it is also helpful to remove outliers where possible. Also check your data for multicollearity; closely related variables are weighted more heavily.

The current sample is from 25 European countries before the fall of the Berlin Wall. The average amount of proteins consumed per capita is categorized into 9 different food groups, eggs, nuts, starch and so on. We are interested in finding out whether or not some countries share similar patterns of food consumption. For those not agriculturally inclined, my main point in using this example is its clarity; when you study a subject of your own choosing, you can enjoy as many complications as you like!

Country	RedMeat	WhiteMeat	Eggs	Milk	Fish	Cereals	Starch	Nuts	FruitVeg
Albania	10.1	1.4	0.5	8.9	0.2	42.3	0.6	5.5	1.7
Austria	8.9	14.0	4.3	19.9	2.1	28.0	3.6	1.3	4.3
Belgium	13.5	9.3	4.1	17.5	4.5	26.6	5.7	2.1	4.0
Bulgaria	7.8	6.0	1.6	8.3	1.2	56.7	1.1	3.7	4.2
Czechoslovakia	9.7	11.4	2.8	12.5	2.0	34.3	5.0	1.1	4.0
Denmark	10.6	10.8	3.7	25.0	9.9	21.9	4.8	0.7	2.4
E Germany	8.4	11.6	3.7	11.1	5.4	24.6	6.5	0.8	3.6
Finland	9.5	4.9	2.7	33.7	5.8	26.3	5.1	1.0	1.4
France	18.0	9.9	3.3	19.5	5.7	28.1	4.8	2.4	6.5
Greece	10.2	3.0	2.8	17.6	5.9	41.7	2.2	7.8	6.5
Hungary	5.3	12.4	2.9	9.7	0.3	40.1	4.0	5.4	4.2
Ireland	13.9	10.0	4.7	25.8	2.2	24.0	6.2	1.6	2.9
Italy	9.0	5.1	2.9	13.7	3.4	36.8	2.1	4.3	6.7
Netherlands	9.5	13.6	3.6	23.4	2.5	22.4	4.2	1.8	3.7
Norway	9.4	4.7	2.7	23.3	9.7	23.0	4.6	1.6	2.7
Poland	6.9	10.2	2.7	19.3	3.0	36.1	5.9	2.0	6.6
Portugal	6.2	3.7	1.1	4.9	14.2	27.0	5.9	4.7	7.9
Romania	6.2	6.3	1.5	11.1	1.0	49.6	3.1	5.3	2.8
Spain	7.1	3.4	3.1	8.6	7.0	29.2	5.7	5.9	7.2
Sweden	9.9	7.8	3.5	24.7	7.5	19.5	3.7	1.4	2.0
Switzerland	13.1	10.1	3.1	23.8	2.3	25.6	2.8	2.4	4.9
UK	17.4	5.7	4.7	20.6	4.3	24.3	4.7	3.4	3.3
USSR	9.3	4.6	2.1	16.6	3.0	43.6	6.4	3.4	2.9
W Germany	11.4	12.5	4.1	18.8	3.4	18.6	5.2	1.5	3.8
Yugoslavia	4.4	5.0	1.2	9.5	0.6	55.9	3.0	5.7	3.2

Open the file Food.csv in Jamovi. To install the **snowCluster** module, go to Modules on the top right and open the jamovi library.

snowCluster

Press the snowCluster tab. From the 'Cluster' sub-menu, select 'Hierarchical Clustering Dendrogram'.

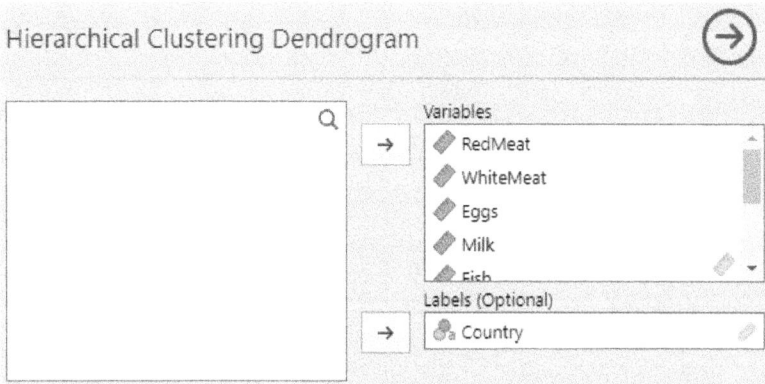

'Country' goes into 'Labels', all the food measures into 'Variables'.

Cluster Dendrogram

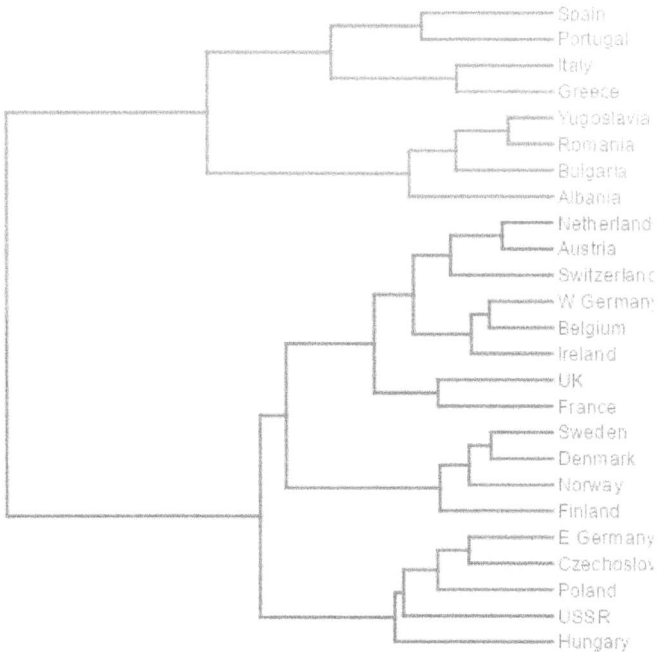

The default dendrogram shows two sets of clusters (in glorious color, but not here), perhaps southern European countries versus the rest. For alternative rationales, we can go to the Options section and increase the number of clusters.

A three-cluster dendogram breaks up Southern Europe into the Balkans and the Mediterrean peninsulae.

A four-cluster configuration offers more than one rationale. A new grouping comprised of East Germany, Czechoslovakia, Poland, USSR and Hungary is a region encompassing Central and Eastern Europe. Another explanation is a political one: the countries making up the new cluster, together with those in the Balkans, were run by Soviet or other communist régimes; the Mediterranean and more northerly states were non-Communist.

If you try again with five clusters, the Scandinavian countries become a separate cluster. With six clusters, the Mediterranean splits into the Iberian peninsual and more easterly states. A seven-cluster configuration produces a UK-France splinter. Clearly it is for the informed investigator to work out which of these makes most sense.

Within the Options section of the Hierarchical Clustering Diagram, there are neat alternative plot types, circular and phylogenic.

For an alternative visualisation, go to snowCluster's 'Cluster' sub-menu, and   select 'Hierarchical clustering method'. This offers a vertical dendrogram by default, with light red boxes to separate the clusters. Also available with this tool are a heatmap and plotted pairs of variables.

It may be worth experimenting with different combinations of distance measures and clustering methods. These may lend themselves to different interpretations; as cluster analysis is an exploratory method, this is not necessarily 'dredging'. Having said that, some choices are to be preferred in certain contexts; below, we consider the available individual options.

Distance measures

The metrics measure the distance between each individual observation.

euclidean – the 'Euclidean distance' option, also known as *L2* and *squared Euclidean distance* – is generally used with the centroid and Ward's clustering methods. This is a widely used distance measure, particularly suitable for interval data.

manhattan – the 'Manhattan distance', also known as *city block* and *L1* – is similar to Euclidean distance, but assumes non-correlation and is useful for damping down outliers.

maximum – the 'Maximum' option, also known as the *Chebishev distance* – measures the greatest differences between vectors. This is useful if you want to consider any objects as 'different' if they differ on any single dimension.

canberra – 'Canberra' - for ordinal data.

binary – 'binary' – for categorical data.

minkowski – 'Minkowski' – particularly good for when interval data has an absolute zero.

Clustering methods

Hierarchical methods "attempt to maximize the differences between clusters relative to the variation within the clusters" (Hair and Black 2000). Most of these, including the options cited here, are called agglomerative procedures, where the clusters are formed by building up on the original objects. 'Divisive procedures' begin with a large cluster and remove dissimilar clusters. The following are the clustering, or 'linkage', methods available in snowCluster:

single – the 'single linkage' method – uses the 'nearest neighbour' or 'friends of friends' approach. If clusters are poorly delineated, 'snake chains' (that's what they look like) may join dissimilar clusters.

complete – the 'complete linkage' option – uses the 'furthest-neighbour approach', examining the furthest cluster elements, to find similar clusters. This eliminates snaking. The other methods are compromises between this and the single linkage option.

average – the 'average linkage' option, also known as *UPGMA* – is initially similar to single and complete linkage methods, but compares the average of individuals in one cluster against all individuals in another. This tends to combine where there is little within-cluster variation. The bias is where clusters have a similar variance.

ward.D2 – 'Ward's method' – is popular in that it generally provides a small number of compact clusters. It is biased towards clusters with the same number of observations.

ward.D – 'Ward.D' – is an alternative version of Ward's distance and should only be used with the Euclidean distance measure (see above). Ward.D2 is the original method and is generally preferred.

centroid – 'centroid method', otherwise known as *UPGMC* – uses the distance between the centroids (the mean value for each variable) and then computes new centroids. This is popular in biology but can be messy and confusing.

mcquitty – 'McQuitty' – also known as *WPGMA* – was evaluated as a particularly effective clustering method by Gavioli *et al* (2019) in the context of management zones in precision agriculture It is similar to UPGMA (average linkage): WPGMA is weighted, UPGMA unweighted.

median – 'median' – also known as *WPGMC*, this is apparently based upon centroids (not medians).

The above descriptions should give you somewhere to start in particular contexts, but these are not hard and fast recommendations: ".. it is advisable to use several measures and compare the results with theoretical or known patterns" (Hair and Black 2000). Unlike many of the methods we have seen, this is not an inference method and has no gold standard of objectivity. Indeed, even the idea of what constitutes a cluster is rather subjective (Kendall 1975), so in this area of research it is perfectly acceptable to dredge for the best outcomes.

k-means clustering

k-means clustering is created using centroid-based algorithms. Located within snowCluster's 'Cluster' sub-menu, it is set up similarly to the hierarchical methods. However, there is no direct link to the labels, so clusters need to be identified by reference to their individual elements. Here, four clusters have been chosen.

Cluster plot

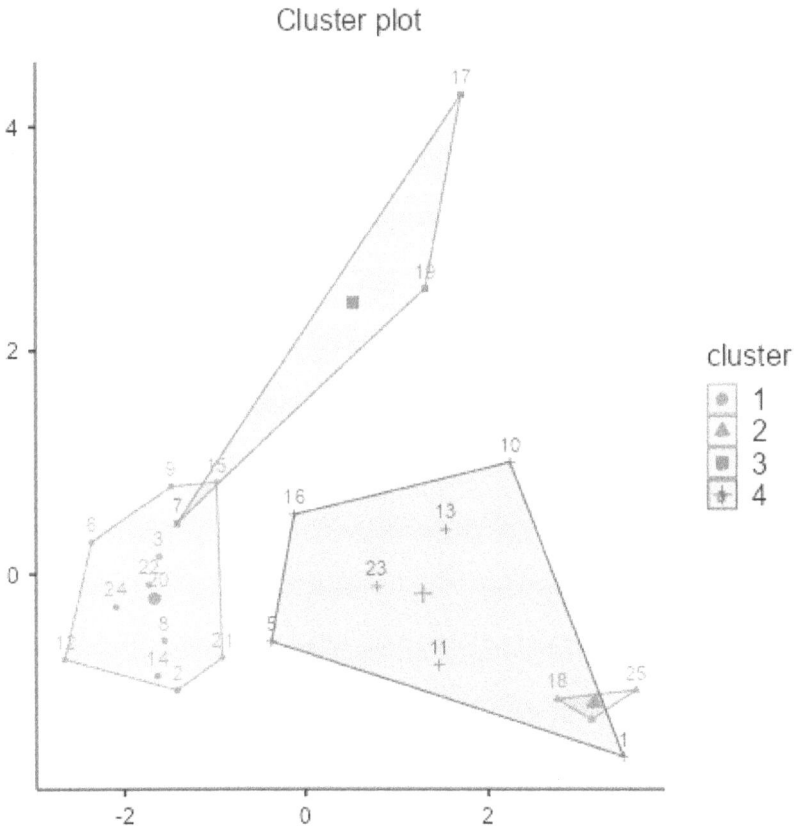

The cluster plot shows the relative positions between different elements. *k*-means may come in useful for fine-tuning the initial findings of hierarchical cluster analysis.

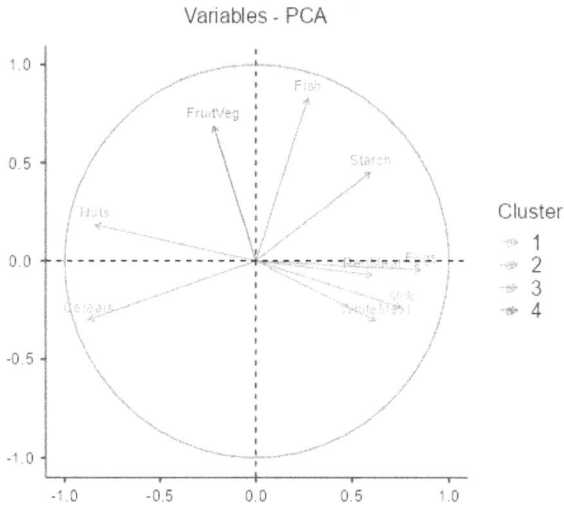

Variables - PCA

The principal components analysis (PCA) of the variables may also be instructive. See next, however, the possibility of 'scientificating' what should really be an aid to judgement ('Optimal number of clusters'):

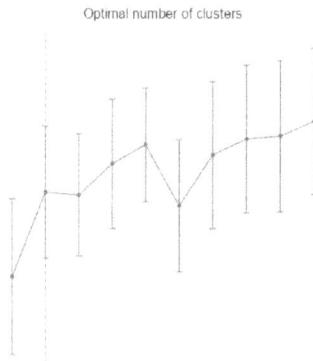

Optimal number of clusters

We are told that two clusters are optimal. Well, maybe in some test tube samples, but not necessarily in our example.

To objectively examine emergent cluster analysis groupings, you may want to classify your findings with a test such as logistic regression.

References

Aldrich JH and Nelson FD (1984) *Linear probability, logit, and probit models.* Beverley Hills, CA: Sage.

Alva JAV and Estrada EG (2009) A Generalization of Shapiro-Wilk's Test for Multivariate Normality. *Communications in Statistics, 38,* 1870-1883.

Bross IDJ (1971) Critical Levels, Statistical Language and Scientific Inference. In Godambe VP and Sprott (eds) *Foundations of Statistical Inference.* Toronto: Holt, Rinehart & Winston of Canada, Ltd.

Bryant FB and Yarnold PR (1995) Principal-Components Analysis and Exploratory and Confirmatory Factor Analysis. In Grimm LG and Yarnold PR (eds) *Reading and Understanding Multivariate Statistics.* Washington: American Psychological Association.

Buser K (1995) Dangers in using ANCOVA to evaluate special education program effects. At *Annual meeting of the American Educational Research Association.* 18–22 April, San Francisco, CA. Educational Resources Center, <www.eric.ed.gov>

Calkins KG (2005) *An Introduction to Statistics* <https://www.andrews.edu/~ calkins/math/edrm611/edrm05.htm>

Campbell K (1989) Dangers in using analysis of covariance procedures. At *Annual Meeting of the Mid–South Educational Research Association* (17th, Louisville, KY, Nov 9–11, 1988), ERIC (Educational Resources Center), www.eric.ed.gov

Clark–Carter D (1997) *Doing quantitative psychological research: from design to report.* Hove: Psychology Press.

311

Cliff N (1987) *Analyzing multivariate data.* San Diego: Harcourt Brace Jovanovich.

Cohen J (1977). *Statistical power analysis for the behavioral sciencies.* Routledge.

Costello AB and Osborne JW (2005) Best Practices in Exploratory Factor Analysis: Four Recommendations for Getting the Most From Your Analysis. *Practical Assessment, Research & Evaluation, 10* (7). <http://pareonline.net/getvn.asp?v=10&n=7>

Dallal GE (2012) *Multiple Comparison Procedures.* <www.jerrydallal.com/lhsp/mc.htm>

Darboux JG, Appell PE and Poincaré JH (1908) Examen critique des divers systèmes ou études graphologiques auxquels a donné lieu le bordereau. In *L'affaire Drefus - La révision du procès de Rennes - enquête de la chambre criminelle de la Cour de Cassation.* Ligue francaise des droits de l'homme et du citoyen, Paris, 499-600.

Dennis B (1996) Discussion: Should Ecologists Become Bayesians? *Ecological Applications, 6,* 1095-1103. <http://www.webpages.uidaho.edu/~ brian/reprints /Dennis_Ecological_Applications_1996.pdf>

Dinno A (2009) Exploring the Sensitivity of Horn's Parallel Analysis to the Distributional Form of Random Data. *Multivariate Behavioral Research, 44,* 362–388.

Everitt B, Landau, S, Leese, M and Stahl, D (2011) *Cluster Analysis.* Oxford: Wiley-Blackwell.

Fay MP (2015) *Exact McNemar's Test and Matching Confidence Intervals* < https://cran.r-project.org/web/packages/exact2x2/vignettes/exactMcNemar.pdf>

Fisher RA (1926), The Arrangement of Field Experiments, *Journal of the Ministry of Agriculture of Great Britain, 33,* 503-513.

Fisher, RA (1935) *The design of experiments.* Edinburgh: Oliver and Boyd.

Gavioli A, de Souza EG, Bazzi CL, Schenatto K and Betzek NM (2019) Identification of management zones in precision agriculture: An evaluation of alternative cluster analysis methods. *Biosystems Engineering, 181*, 86-102.

Gelman A (2011). Induction and Deduction in Bayesian Data Analysis. *Rationality, Markets and Morals, 2*(43). < http://wiki.learnstream.org/wiki/ref:gelman2011induction>

Gelman A and Rubin DB (1995) Avoiding model selection in Bayesian social research. In Marsden PV (ed) *Sociological Methodology, 25*.

Girden E (1992). *ANOVA: Repeated measures*. Newbury Park, CA: Sage.

Gorsuch RL (1983) *Factor Analysis*. Hillsdale NJ: Lawrence Erlbaum.

Greene J and D'Oliveira M (1982). *Learning to use statistical tests in psychology: A student's guide*. Milton Keynes: Open University Press.

Griffin L (2007) Historical sociology, narrative and event-structure analysis: fifteen years later. *Sociologica, 3*, 1-17.

Hair JF and Black WC (2000) Cluster Analysis, in Grimm LG and Yarnold PR (eds) *Reading and Understanding More Multivariate Statistics*. Washington DC: American Psychological Association.

Hajian-Tilaki K (2013) Receiver Operating Characteristic (ROC) Curve Analysis for Medical Diagnostic Test Evaluation. *Caspian Journal of Internal Medicine, 4*, 627–635.

Hilton A and Armstrong RA (2006) Statnote 6: post-hoc ANOVA tests. *Microbiologist, 7*, 34-36.

Holm S (1979) A simple sequential rejective multiple test procedure. *Scandinavian Journal of Statistics, 6*, 65-70.

Horn JL (1977) Personal communication.

Hosmer DW and Lemeshow S (2000) *Applied Logistic Regression*, 2nd Edition. New York: Wiley.

Hsu JC (1996) *Multiple Comparisons: Theory and Methods*. London: Chapman and Hall.

Huck, SW (2008) *Statistical misconceptions*. London: Routledge.

Huck SW (2012) *Reading Statistics and Research.* Boston: Pearson.

Jarosz AF and Wiley J (2014) What Are the Odds? A Practical Guide to Computing and Reporting Bayes Factors. *Journal of Problem Solving, 7.*

Jeffreys H (1961) *Theory of Probability* (3rd edition). Oxford: Clarendon Press.

Kahneman D (2011) *Thinking, Fast and Slow.* London: Allen Lane.

Karadeniz PG and Ercan I (2017) Examining tests for comparing survival curves with right censored data. *Statistics in Transition, 18,* 311-328.

Kass RE and Raftery AE (1995) Bayes Factors. *Journal of the American Statistical Association, 90,* 773-795.

Katani K (2014) *Koji Yatani's Course Webpage* < http://yatani.jp/teaching/doku.php?id=hcistats:chisquare >

Kendall M (1975) *Multivariate Analysis.* London: Griffin.

Kenny DA (2015) *Measuring Model Fit.* <davidakenny.net/cm/fit.htm>

Kieffer K M (1998) Orthogonal versus Oblique Factor Rotation: A Review of the Literature regarding the Pros and Cons. At *Annual Meeting of the Mid-South Educational Research Association*, New Orleans, LA.

Kiers HAL (1994) Simplimax: Oblique rotation to an optimal target with simple structure. *Psychometrika, 59,* 567-579.

Kim HY (2013) Statistical notes for clinical researchers: assessing normal distribution (2) using skewness and kurtosis. *Restorative Dentistry and Endedontics, 38,* 52-54. <http://www.ncbi.nlm.nih.gov/pmc/articles/PMC3591587/>

Kim S (2015) ppcor: An R Package for a Fast Calculation to Semi-partial Correlation Coefficients. *Communications for Statistical Applications and Methods, 22,* 665-674.

Kinnear P and Gray C (2004) *SPSS 12 made simple.* Hove: Psychology Press.

Kinnear P and Gray C (2008) *SPSS 15 made simple.* Hove: Psychology Press.

Kucharski A (2016) How science and statistics are taking over sport. *New Statesman*, 30 April 2016.

Ladesma RD and Valera-Mora P (2007) Determining the Number of Factors to Retain in EFA: an easy-to-use computer program for carrying out Parallel Analysis. *Practical Assessment, Research & Evaluation, 12*, 1-11.

Lee MD and Wagenmakers E-J (2013). *Bayesian modeling for cognitive science: A practical course.* Cambridge University Press.

Lewin K, Lippitt R, and White RK (1939). Patterns of aggressive behavior in experimentally created "social climates." *The Journal of Social Psychology, 10*, 271-299.

Little RJA (1978) Generalized Linear Models for Cross-Classified Data from the WFS. *World Fertility Survey Technical Bulletins, 5*.

McFadden D (1979) Quantitative Methods for Analyzing Travel Behaviour on Individuals: Some Recent Developments. In Hensher D and Stopher P (eds) *Behavioural Travel Modelling.* London: Croom Helm.

McGrayne SB (2012) *The theory that would not die: how Bayes' Rule cracked the Enigma Code, hunted down Russian submarines, and emerged triumphant from two centuries of controversy.* New Haven CT: Yale University Press.

Madsen M (1976). Statistical Analysis of Multiple Contingency Tables. Two Examples. *Scandinavian Journal of Statistics,3*, 97-106.

Mayo D (2012) <http://errorstatistics.blogspot.co.uk/#uds-search-results>

Mead R, Gilmour SJ and Mead A (2012) *Statistical Principles for the Design of Experiments.* Cambridge: Cambridge University Press.

Miller G and Chapman J (2001) Misunderstanding analysis of covariance. *Journal of Abnormal Psychology 110*(1), 40–8.

Morey RD, Hoekstra R, Rouder JN, Lee MD and Wagenmakers E-J (2016) The Fallacy of Placing Confidence in Confidence Intervals. *Psychonomic Bulletin & Review, 23*, 103-123.

Mortensen U (1974) Bayesian sequential analysis in Psychological Research *New Zealand Journal of Psychology*, *3*(1), 37-44 http://www.psychology.org.nz/wp-content/uploads/PSYCH-Vol31-1974-6-Mortensen.pdf

Nelder J (1971) Discussion. *Journal of the Royal Statistical Society*, *series B*, *33*, pp 244-246.

Nelder J (1999) From statistics to statistical science. *Statistician*, *48*, 257–267.

New Zealand Ministry of Education (2016) <http://www.wellbeingatschool.org.nz/information-sheet/understanding-and-interpreting-box-plots>

Okada K (2013) Is omega squared less biased? A comparison of three major effect size indices in one-way ANOVA. *Behaviormetrika*, *40*, 129–147.

Parker R (1979) *Introductory statistics for biology*. London: Edward Arnold.

Pedhazur EJ and Schmelkin LP (1991) *Measurement, Design and Analysis*. Hillsdale NJ: Lawrence Erlbaum.

Perrigot R, Cliquet G and Mesbah M(2004) Possible applications of survival analysis in franchising research. *International Review of Retail, Distribution and Consumer Research*, *14*, 129-143. <http://www.lsta.upmc.fr/mesbah/42)%20Perrigot,%20R., %20Cliquet,%20G.%20and%20Mesbah, %20M.%20(2004).pdf >

Pickren WE and Rutherford A (2010) *A History of Modern Psychology in Context*. New York: John Wiley.

Plackett, R. (1971) *Introduction to the theory of statistics*. Edinburgh: Oliver and Boyd.

Popper K (1968) *The Logic of Scientific Discovery*. New York: Harper and Row.

Preece, D. (1982) T is for trouble (and textbooks): a critique of some examples of the paired-samples t–test. *The Statistician*, *31*, 169–195.

Raftery AE (1995). Bayesian model selection in social research. In Marsden PV (ed), *Sociological methodology*, 111–196. Cambridge, MA: Blackwell.

Razali NM and Wah YB (2011). Power Comparisons of Shapiro-Wilk, Kolmogorov-Smirnov, Lilliefors and Anderson-Darling Tests. *Journal of Statistical Modeling and Anlytics*, *2*, 21-33.

Rennie KM (1997) Exploratory and Confirmatory Rotation Strategies in Exploratory Factor Analysis. At *Annual meeting of the Southwest Educational Research Association*, Austin TX.

Revelle W (2016) Personal communication with the author.

Reyment RA, Blackith RE and Campbell NA (1984) *Multivariate morphometrics*. New York: Academic Press.

Rice W (1989) Analysing tables of statistical tests. *Evolution*, *43*, 223–225.

Schonbrodt F (2015) *What does a Bayes Factor feel like?* <http://www.nicebread.de/what-does-a-bayes-factor-feel-like/>

Sezen-Balcikanli G and Sezen M (2017) Professional sports and empathy: relationships between professional futsal players' tendency toward empathy and fouls. *Physical Culture and Sport: Studies and Research*, *73*, 27-35.

Smolinski L (1973) Karl Marx and Mathematical Economics. *Journal of Political Economy*, *81*, 1189-1204.

StatsDirect statistical software Version 2 (2011).

Steiger JH (1979) Factor indeterminacy in the 1930s and the 1970s: Some interesting parallels. *Psychometrika*, *44*, 157-167. Cited in Stevens (2009).

Steinhardt J (2014) *A Fervent Defense of Frequentist Statistics.* <http://lesswrong.com /lw/jne/a_fervent_defense_of_frequentist_statistics/>

Stevens JP (1996) *Applied Multivariate Statistics for the Social Sciences* (3rd ed). Mahwah, NJ: Erlbaum.

Stevens JP (2009) *Applied Multivariate Statistics for the Social Sciences* (5th ed). NY: Routledge.

Tabachnick BG and Fidell LS (2007) *Using Multivariate Statistics.* Boston: Pearson.

Tabachnick BG and Fidell LS (2013) *Using Multivariate Statistics. Boston: Pearson.*

Thomson A and Randall-Maciver R (1905) *Ancient Races of the Thebaid.* Oxford: Oxford University Press.

Tsoumakas G, Lefteris A and Vlahavas I (2005) Selective fusion of heterogeneous classifiers. *Selective Data Analysis, 9*, 511–525.

Veall MR and Zimmermann KF (1996) Pseudo-R^2 Measures for Some Common Limited Dependent Variable Models. *Journal of Economic Surveys, 10* (3), 241-259.

Vittinghoff E and McCulloch CE (2007) Relaxing the Rule of Ten Events per Variable in Logistic and Cox Regression. *American Journal of Epidemiology, 165*, 710-718. <http://aje.oxfordjournals.org/content/165/6/710.full>

Wagenmakers E (2007) A practical solution to the pervasive problems of p values. *Psychonomic Bulletin & Review, 14*(5), 779-804 < http://www.ejwagenmakers.com/2007/pValueProblems.pdf>

Wanless E, Jensen J and Poliakoff P (2019) New applications of survival analysis modeling: Examining intercollegiate athletic donor relationship dissolution. *Journal of Sports Analytics, 5*, 45-56.

Weinfurt KP (1995) Multivariate Analysis of Variance. In Grimm LG and Yarnold PR (eds) *Reading and Understanding Multivariate Statistics.* Washington: American Psychological Association.

West SG, Finch JF, Curran PJ (1995) Structural equation models with nonnormal variables: problems and remedies. In Hoyle RH (ed) *Structural equation modeling: Concepts, issues and applications*, 56–75. Newbery Park, CA: Sage.

Wilner D, Walkley RR and Cook SW (1955) *Human relations in interracial housing: A study of the contact hypothesis.* Minneapolis: University of Minnesota Press.

Wright RE (1995) Logistic Regression. In Grimm LG and Yarnold PR (eds) *Reading and Understanding Multivariate Statistics.* Washington: American Psychological Association.

Wyseure G (2003) *Multiple comparisons'*
 http://www.agr.kuleuven.ac.be/vakken/statisticsbyR
 /ANOVAbyRr/multiplecomp.htm

Yap BW and Sim CH (2011) Comparisons of various types of
 normality tests. *Journal of Statistical Computation and Simulation*,
 81, 2141-2155.

Yerkes RM and Dodson JD (1908) The Relation of Strength of Stimulus
 to Rapidity of Habit Formation. *Journal of Comparative Neurology
 and Psychology*, *18*, 459–482.
 <https://doi.org/10.1002/cne.920180503>

Index

confidence intervals, 28, 56-7, 75, 76, 215, 235-6, 239, 248,
    250, 254-5, 296
corrections, see multiple comparisons
Cohen's d, 31, 56, 73
Cramer's V, 124-5
data types, 16, 21-2
degrees of freedom (df), 124, 148
dispersion, 20, 22, 45, 75-6, 150
Dwass test, 65
effect size, 27, 31, 43-4, 49, 56, 61-2, 65, 73, 75, 77, 81, 82, 86,
    91, 96, 103, 124-26, 147, 151, 153, 158, 162, 166, 201,
    241, 265-6, 275, 279, 282
equality of variances, see homogeneity of variance
estimated marginal means, 240
eta squared, 49, 61-2
factor analysis and PCA, 194-226
    Bartlett's test of sphericity, 198, 202-3
    BIC (Schwarz) criterion, 216, 218-19
    confirmatory factor analysis, 194, 195
    eigenvalues, 196, 206-209, 212, 222
    exploratory factor analysis, 194, 195, 195-8, 212-19
    Horn's Parallel Analysis (PA), 198, 198-9, 200
    Kaiser's criterion (Guttman-Kaiser), 199, 206-09, 216, 219-20
    KMO / MSA (sampling adequacy), 198, 202-5
    loadings, 196, 201, 202, 205, 214-5, 217, 218, 222, 223
    principal components analysis, 194-212, 219-226
    RMSEA, 215
    rotation (oblique and orthogonal), 196-7, 219, 222-25
    scree plot, 199, 206-8, 211-2, 216, 220
    simple structure, 196-7, 211, 224, 225
    TLI (Tucker Lewis Index or NNFI), 215
F ratio, 48-9, 61, 94, 103, 162
Friedman test, 50-52
gamma, 125-6
Greenhouse-Geisser and Huyn-Feldt corrections, 174-5

Mann-Whitney test, 36, 55, 57-59
MANOVA (multivariate analysis of variance), 184-191
    Box's M test, 186, 187
    Hotelling, Roy, Pillai, Wilks, 188-9
Mauchly's W (test of sphericity), 48, 174-5
McNemar test, 135-142
multinomial (chi square goodness of fit) test, 114-18
multiple / pairwise comparison tests, 176-9
    Bonferroni correction, 49, 175, 178, 190, 300
    Holm, 162, 175, 178, 190, 300
    Scheffe, 178
    Tukey, 62-3, 171, 178
multiple correlations, 82-86
multiple regression, 95-105
    hierarchical, 101-105
    standard (simultaneous), 95-101
non-parametric tests, 20-22, 31, 37, 42, 43, 44, 50, 58, 61, 66, 149, 197, 282, 290, 300
normal (Gaussian) distribution, 22, 23, 31, 37, 41, 45, 58, 74, 77, 85, 149, 170, 185, 187, 188
Omega squared, 60-62
outliers / outlying data, 20, 74, 93, 151, 198, 303, 307
overfitting (of models), 218-19, 223
parametric tests, 20-23, 37, 43, 44
parsimonious model, 100, 103, 134, 216, 218-19, 223, 225, 235, 238, 241
part correlations, see semi-partial correlations
partial correlations, 257-60
partial eta squared, 49, 61-2, 162
Pearson test, 69-70, 71, 73, 73-6, 82-4
phi, 124
population, 22, 24, 29, 264-5
prior distribution, 270-72
post hoc tests, see multiple comparisons
Principal Components Analysis (PCA), see factor analysis

prior distribution, 277-9
qualitative analysis, see research design
range, 20, 75-6
research design, 13-16, 35-8
    categorical (nominal, qualitative), 16, 107-109
    control group, 13
    correlational, 15, 67-8
    dependent variable and criterion, 13-14, 38, 88
    experimental, 13, 37-8
    independent variable and predictor, 13-14, 38, 88
    pairing (matching), 14-15, 36
    quasi–experimental, 14, 38
    related design (same subjects), 14-15, 35-6
    time series (longitudinal), 44
    unrelated design (different subjects), 14-15, 35-6
residuals, 61, 97-8
RMSE (in multiple regression), 100, 103, 104
samples, 22, 23, 24, 27, 29, 74, 77, 99, 105, 123-4, 142, 185, 198-9, 208, 211-12, 215, 220, 229, 264-5, 281, 282, 303
semi-partial correlations, 260-61
Shapiro-Wilk test, 23, 30-31, 40-41, 45, 74, 77, 85, 185
significance, 24-32
    alternative hypothesis, 25
    critical value (or 'alpha'), 26-7, 42, 43, 49, 72, 79, 84, 112, 123, 142, 150, 186, 188, 190, 265, 274, 279, 282, 300
    null hypothesis, 25
    one-tailed hypothesis, 28-9
    p value, 25-26, 28-9, 69-71, 124, 150
    two-tailed hypothesis, 28-9
    type one and type two errors, 27, 29, 46-7, 63, 82, 124, 161, 174, 176, 178, 184, 190, 299
skewness, 23
Spearman, 71, 74, 77-9, 85-6
sphericity, 48, 161, 174-5
standard deviation, 20, 22, 31, 56, 62, 76, 150

www.ingramcontent.com/pod-product-compliance
Lightning Source LLC
Chambersburg PA
CBHW052127030426
42337CB00028B/5053